MATLAB® and Design Recipes for Earth Sciences

Martin H. Trauth · Elisabeth Sillmann

MATLAB® and Design Recipes for Earth Sciences

How to Collect, Process and Present Geoscientific Information

 Springer

Martin H. Trauth
Inst. für Erd- und Umweltwissenschaften
Universität Potsdam
Potsdam
Germany

Elisabeth Sillmann
BlaetterwaldDesign
Landau
Germany

Additional material to this book can be downloaded from http://extras.springer.com/

ISBN 978-3-642-32543-4   ISBN 978-3-642-32544-1   (eBook)
DOI 10.1007/978-3-642-32544-1
Springer Heidelberg New York Dordrecht London

Library of Congress Control Number: 2012945735

Printed on acid-free paper

Springer is part of Springer Science+Business Media (www.springer.com)

Preface

The book *MATLAB and Design Recipes for Earth Sciences* is designed to help undergraduate and postgraduate students, doctoral students, post-doctoral researchers, and professionals find quick solutions for common problems when starting out on a new research project. A project usually starts with searching and reviewing the relevant literature and data, and then extracting relevant information as text, data or graphs from the literature, followed by searching, processing and visualizing the data, and finally, compiling and presenting the results as posters, abstracts and talks at conferences.

The course on which this book is based was first taught by M.H.T. as a bachelor's module for second-year students during the 2010/11 winter semester, three years after the introduction of bachelor's and master's programs at the University of Potsdam. The initial design of the bachelor's program included an introductory course on data analysis, scheduled for the second year, which was based on the sister book to this one: *MATLAB Recipes for Earth Sciences–3rd Edition* (Trauth 2010). This course was a complete failure, probably because the second-year bachelor students were not well enough prepared for an advanced course on data analysis, even after two semesters of mathematics during the first year. A few weeks later, the course for students at master's and doctoral levels on the same topic, which M.H.T. was invited to give at the University of Ghent in Belgium, was a great success. The difference between the undergraduate students in Potsdam and the graduate students in Belgium was, of course, the greater motivation that students already working on their own projects had to learn the statistical and numerical methods offered by MATLAB, in order to be able to analyze their data.

As a consequence, M.H.T. moved the course into the master's program and designed a completely new course on *How to Collect, Process and Present Geoscientific Information*, which was very well received by the second-year students, despite the very large number of participants. The course was not presented as a complete package, but evolved during the months of teaching, taking into consideration the suggestions made by students attending the course. During the course, and very much motivated by its success, the idea for this new book quickly emerged and the first outline for the text was drafted in late December 2010. Most of the text was written immediately following completion of the first course and before the start of the second course in spring 2011. Fortunately, the graphic design specialist E.S., who is the owner of *blaetterwaldDesign*, joined the project to contribute to the design sections in the book as well as the book's layout, after having designed the layout of all three editions of the sister book, as well as many other books, for *Springer*. The publisher quickly agreed to assist in realizing the book and contracts were signed in summer 2011.

While undergraduates participating in a course on data analysis might wish to work their way through the entire book, more experienced readers might refer to only one particular method in the book, in order to solve a specific problem. The concept of the book and its contents are therefore outlined below, in order to make it easier for readers with a variety of different requirements to decide how they wish to approach the book.

- *Chapter 1* – This chapter is about initiating, planning and organizing a project. It introduces the Internet resources used in the following chapters to search for geoscientific information, as well as the software and online tools used to manage projects, to process data, to exchange information, and to present project results.

- *Chapters 2 and 3* – These chapters deal with searching and reviewing scientific literature and data on the Internet. Chapter 2 provides a comprehensive tutorial-style introduction to Internet literature resources. It also demonstrates how to extract information from the literature for use within the reader's own projects, and introduces software for managing large collections of electronic journal articles and books. Chapter 3 introduces the most popular data formats on the Internet, and methods to store and transfer such data. Data access and management is demonstrated by means of typical examples.

- *Chapters 4 to 7* – The first of these chapters starts with a tutorial-style introduction to MATLAB, designed for earth scientists (as in Chapter 2 of the sister book). Chapters 5 and 6 introduce advanced visualization techniques with MATLAB, for example, how to create sophisticated two- and three-dimensional graphs from data collected in Chapter 3. Chapter 7 is on processing and displaying images with MATLAB, including satellite images (as in Chapter 8 of the sister book).

- *Chapter 8* – The graphs created with MATLAB in the previous chapters are now handed over to the graphic design unit of the project. Even though the advanced plotting features of MATLAB presented in Chapters 5 and 6 are able to create sophisticated figures, all graphs will require further editing with vector and image processing software before they can be included, together with text and tables, in conference presentations and manuscripts.

- *Chapters 9 to 11* – These chapters are about creating conference presentations such as talks and posters, and various types of manuscripts for publication. They cover the preparation of colorful flyers and brochures relating to projects, as well as theses or project reports with relatively modest designs, and also deal with assembling books and their layout design. Both Chapters 9 and 10 close with some remarks on practicing for conference presentations, and their final delivery.

The book contains *MATLAB* scripts, or *M-files* for visualizing typical earth science data sets (http://mathworks.com). The MATLAB codes can be easily adapted to the reader's data and projects. M.H.T. developed these recipes using MATLAB Version 7 (R2011b), but most of them will also work with earlier software releases. Furthermore, the book relies on numerous other software products, first and foremost the *Adobe Creative Suite* (http://adobe.com), which is used to edit all the graphs created with MATLAB. Although most examples are also explained with open-source alternatives, the use of the Adobe Creative Suite produces consistently high quality results for all graphics to be included in project presentations. The book provides brief introductions to the use of these graphics editors by means of step-by-step tutorials, supplemented by screenshots documenting the workflows that are provided as supplementary electronic material to this book. We are planning to make all vector materials available online as soon as an appropriate digital rights management is provided by the publisher.

We hope that our readers will appreciate our efforts to introduce open-source software tools in addition to the commercial products that the authors of this book use during their daily work. During the course at the University of Potsdam, students asked about free alternatives to *MATLAB*, such as *Python*, *R* and *Octave*. Some students also liked to use *LaTeX* for typesetting, and *GMT* for creating *xy* and *xyz* plots. Students' financial resources are often limited and many therefore use open-source software on their computers. For professionals, however, time is by far the more important limiting factor. When trying to meet a strict deadline for the submission of a research proposal or report, one quickly learns to appreciate complete and concise software manuals and the short response time of the software vendor's support line.

In putting together this book we have benefited from the comments offered by many people, in particular Nina Bösche, Verena Förster, Oliver Korup, Oliver Oswald, and Marius Walter. It is expected that this book will be constantly changing and evolving over time, as has been the case through the various editions of its sister book. Please send us your comments and criticisms on the text, suggestions for correction and expansion of the text, and comments on any experiences that you may have had with similar courses or books.

Please visit the webpages of M.H.T. (http://www.geo.uni-potsdam.de/palaeoklimadynamik.html) and E.S. (http://blaetterwald-design.de) from time to time, in order to check for updates and errata files for this book.

We are much obliged to Ed Manning for professional proofreading of the text. We would like to thank Christian Witschel, Chris Bendall and their team at *Springer*, and also Andreas Bohlen, Brunhilde Schulz and their team at *UP Transfer GmbH* for their support. M.H.T. acknowledges the *Book Program* and the *Academic Support* at *The MathWorks Inc.*, as well as Claudia Olrogge, Kremena Radeva, and Annegret Schumann at *The MathWorks GmbH Deutschland*. E.S. thanks *Adobe Systems Inc.* for their support and the permission to include screenshots of Adobe software in the book. M.H.T. would also like to thank *NASA/ GSFC/METI/ERSDAC/ JAROS* and the *U.S./Japan ASTER Science Team*, and their team leader Mike Abrams, for permission to include their ASTER images in the book.

Potsdam/Landau, June 2012

Martin Trauth   *Elisabeth Sillmann*
Earth Scientist, University of Potsdam   Designer (AGD), blaetterwaldDesign.de

apl. Prof. Dr. Martin H. Trauth   Dipl-Ing. (FH) Elisabeth Sillmann
Institute of Earth and Environmental Science   blaetterwaldDesign
University of Potsdam   Büro für Medien und Gestaltung
Karl-Liebknecht-Strasse 24–25   Emich-von-Leiningen-Strasse 38
D-14476 Potsdam   D-76829 Landau in der Pfalz
Germany   Germany

Contents

1 Scientific Information in Earth Sciences

1.1 Introduction

This book is based on an undergraduate course taught at the University of Potsdam in Germany (http://uni-potsdam.de), as was also the case with its sister book *MATLAB Recipes for Earth Sciences–3rd Edition* (Trauth 2010). The objective of this course was to guide students through the typical progression of a scientific project. Such projects usually start with a search of the relevant literature in order to review and rank published books and journal articles, to extract information (as text, data, maps, or graphs), and to search, process and visualize data, compiling the results and presenting them as posters, abstracts and oral presentations (talks). The course was first held for second-year students in earth sciences during the 2010–11 winter semester, and then repeated in the following summer semester. The original plan was to hold the course in a computer pool with fifteen workstations. However, an unexpectedly large number of students enrolled for the first presentation, which had more than sixty participants. This led to the course being held in a lecture hall with a projector, a microphone, and a speaker system. There was also a table for the instructor's laptop and equipment, and wireless access to the Internet; the students used their own private laptops.

The change of teaching rooms had both pros and cons. The large lecture hall of course provided a nice conference-type atmosphere with its audio-video system, light dimming and room darkening systems, and large projection screens. The use of private laptops had the advantage that the students were already familiar with their own computing systems prior to the start of the course. At the end of the course, the participants could carry home their entire workspace and continue the course-related project work at home. On the other hand, a clear disadvantage was that the students' private laptops had hardly any of the required software packages installed, such as *MATLAB*, the *Adobe Creative Suite* and *Microsoft Office*. The obvious solution to this problem was the consistent use of *Open Source Software* instead of commercial products, accepting the clear limitations of most of

M. H. Trauth and E. Sillmann, *MATLAB® and Design Recipes for Earth Sciences*,
DOI: 10.1007/978-3-642-32544-1_1, © Springer-Verlag Berlin Heidelberg 2013

these free software tools such as the often limited range of functions and performance, incomplete documentation and the general lack of support.

The compromise made in the course was to use *MATLAB* commercial software, which is also widely-used in many other courses and is becoming increasingly popular in earth sciences, rather than *Octave*, *Python* or other free products. The *Adobe Creative Suite* (including *InDesign* for desktop publishing, *Illustrator* for editing vector graphics and *Photoshop* for editing pixel graphics) was, however, largely replaced by the free *Scribus*, *Inkscape*, and *Gimp* software. *Microsoft Office* (including *Word* for text processing, *Excel* for spreadsheet calculations and *PowerPoint* for presentations), and the Apple equivalent, *iWork* (including *Pages*, *Numbers* and *Keynote*), were replaced by the free *OpenOffice* suite. Most students purchased the inexpensive *MATLAB Student Version* for their project work, as well as downloading and installing the recommended free software packages. In this book, however, not being a software-oriented tutorial, we use both commercial and free products to demonstrate the principles for collecting, processing, and presenting information in earth sciences.

A major challenge was to handle the large number of participants in a highly interactive, tutorial-style environment. The solution was to establish a reliable eLearning environment for the course using the free *Modular Object-Oriented Dynamic Learning Environment* (Moodle, http://moodle.org), also known as a *Course Management System* (CMS) and widely-used for electronically-supported learning and teaching, including at the University of Potsdam (Fig. 1.1). The Moodle course page provided a section for each weekly class including presentations, teaching manuscripts and additional course materials such as literature, data, example files, and interesting data and information links. The course page also included a message board for announcements, a forum for discussions and drop boxes for homework (Fig. 1.2).

In addition to establishment of the Moodle course environment, the students were organized in three *Google Groups* (http://groups.google.com), not accessed by the course instructor but led by elected student group leaders and their deputies. During the course, the instructor defined three principal topics for each of the groups, within the overall theme of *Past Climate Changes*. One advantage of having the students organized in groups was the introduction of teamwork; collecting literature and data, processing the results, solving minor problems (e.g., technical problems), and organizing the presentations at the end of the course all worked exceptionally well within the groups. The instructor was mainly contacted by the group leaders to answer more fundamental questions and to provide support with organiz-

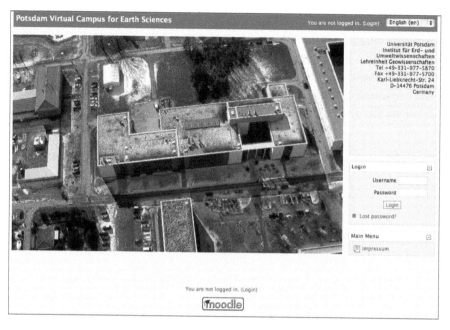

Fig. 1.1 Screenshot of the Potsdam Virtual Campus for Earth Sciences based on the free Modular Object-Oriented Dynamic Learning Environment (Moodle, http://moodle.org), also known as Course Management System (CMS), which is widely-used for electronically-supported learning and teaching, including at the University of Potsdam.

ing and managing the teamwork within the groups.

The following information on the course is provided as it may be useful for instructors intending to design their own course on the basis of this book. The duration of the course at the University of Potsdam was 11 to 13 weeks, depending on the number of participants. There was a three hour class in each of these weeks with lectures, demonstrations, student presentations, and discussions. The typical work load for each student was 8 to 10 hours per week, including the three-hour class; this resulted in 6 credit points under the European Credit Transfer System. The typical syllabus for the courses on *MATLAB and Design Recipes for Earth Science* was as follows:

- Week 1 – Introduction and class on Chapter 1: *Scientific Information in Earth Sciences*. Explanation of the course concept, description of the course examination procedure (including the submission of a 200-word abstract, the design of a poster, and the delivery of a two-minute presentation in the last three weeks of the course), definition of the group's research topics, establishment of groups and election of the group leaders.

- Week 2 – Presentations by group leaders on the establishment and organization of each group (approximately 5 minutes each). Highly interactive class on Chapter 2: *Searching and Reviewing Scientific Literature.*
- Week 3 – Presentations by group leaders of results from their searches for the five most relevant articles on their research topics. Class on Chapter 3: *Internet Resources for Earth Science Data.*

Fig. 1.2 Screenshot of the course webpage providing a section for each weekly class including presentations, teaching manuscripts and additional course materials such as literature, data, example files, interesting data and information links, a message board for announcements, a forum for discussions and drop boxes for homework.

- Week 4 – Reports by group leaders or their deputies (5 to 10 minutes each) on the teamwork within the groups and how they organize and rank the information collected in the groups. Presentation of the 10 to 20 most important articles on their research topics, including a description of the system used to rank these articles and an attempt to display the list in a structured way. Class on Chapter 4: *MATLAB as a Visualization Tool.*
- Weeks 5 to 7 – Reports by each group (one per week, 10–15 minutes each) on the most exciting hypotheses and controversies in their research topics, presenting the relevant references, representative graphs, and data sets. Presentations of summaries by each group with the presentation method being chosen by the group, and with the presentations being made by either a single representative (the group leader, a deputy, or any other group member) or, as some of the groups preferred, by a group of three to five representatives, each presenting different parts of the summary. At this stage of the course the groups start to split their research topic into sub-themes, each of which is pursued by an individual member of the group. Classes on Chapters 5 to 7: *Visualizing Data and Processing and Displaying Images in Earth Sciences.*
- Week 8 – Presentation by each group of an outline for the conference session to be held at the end of the course, followed by discussion of the sub-themes and the proposed titles of the talks. Class on Chapter 8, *Editing Text and Tables* and Chapter 11, *Creating Manuscripts*, since the first student work to be submitted will be the conference abstract.
- Week 9 – Presentation by each group of an updated outline of the conference session. The groups may also present drafts of individual graphs to be included in posters and presentations. Class on Chapter 8, *Editing Vector and Raster Graphics* and Chapter 10, *Creating Conference Posters.* Student examination: each student posts to the Moodle platform a 200-word abstract on their sub-theme.
- Weeks 10 – Presentation by each group of an updated outline of the conference session, and discussion of the schedule for the conference session. The groups can also present drafts of individual conference slides. Class on Chapter 8, *Editing Vector and Raster Graphics* and Chapter 9, *Creating Conference Presentations.* Student examination: each student presents a poster on their sub-theme, including the previously submitted abstract.
- Week 11 – Conference sessions organized by the students. Student examination: each student gives a two-minute presentation in a large lecture hall, in front of a jury of three professors.

Each of these components have combined to form the current design of the course and the concept behind this book. At the University of Potsdam, the course culminated in the presentation of the project work as conference-style posters, 200-word abstracts, and one-hour sessions with 10–15 two-minute presentations, chaired by the project leaders and their deputies. The following sections of this chapter provide a closer look at the design of the course and this companion book, introducing briefly the techniques for collecting, processing and presenting information in earth sciences, as well as the information formats encountered while working with data in our discipline and strategies for optimizing and organizing the workspace and work flow.

1.2   Collecting and Managing Information in Earth Sciences

Initiating, Planning and Organizing a Project

Imagine a small group of people (perhaps two or three) with an idea for a new research project. Having identified a specific topic as a research area worthy of investigation, the initial group starts putting together a team of researchers with the necessary methodological expertise to run the project. After assembling the team, the core group behind the new project writes a research proposal for submission to a funding agency. The proposal is reviewed and, if successful, funding is provided, after which the project can actually start.

During the course, we simulated the initiation, planning, and organization of a research project. In the summer semester course, the instructor defined three research themes as examples of the topics for student projects:

- Tectonics and climate,
- The ice ages during the Pleistocene epoch, and
- Climate through the last millennium.

The instructor then asked the students to designate a team leader for each of the three topics. The students accepted as team leaders then established Google Groups. According to their webpage, Google Groups was started in 2001 and helps project members connect with other members, access information, and communicate effectively by email and on the web. The service provides a mailing list for each group to use for project discussions. Google Groups is linked with *Google Docs* (http://docs.google.com), which became available in 2007 and in which you can create and share text documents, spreadsheets, presentations, and many other types of files so that they can be accessed from anywhere. As an alternative to Google Docs, the students also used *DropBox* (http://dropbox.com), which is a free web-based file-hosting service. DropBox uses a cloud-storage system to synchronize files on different computers with a server via the Internet.

After establishing the groups, the team leaders were then contacted by the remaining students requesting participation in the groups. Each group then selected deputy team leaders and soon after that started forming sub-groups to work on specific aspects or sub-themes of the overall research themes. As an example, the group working on *Tectonics and Climate* as a research topic defined sub-themes such as

- Mountain uplift and climate change
- Closure and opening of ocean gateways and change of climate
- Volcanic activity altering climate on the Earth

The sub-themes were identified by the students themselves soon after they started searching and reviewing the literature on the overall research topic of their group.

Literature and Data Resources on the Internet

The Internet provides fast access to geoscientific information such as conference proceedings, electronic books and journal articles, digital geologic maps, satellite images and earthquake catalogs. Chapters 2 and 3 demonstrate the use of literature search catalogs and various other resources for accessing earth science data.

In 2011 the online book store *Amazon* (http://amazon.com) announced that they had sold more electronic books than printed books. Long before that most researchers and their institutions had switched from printed to electronic collections of journal articles. The advantage of an electronic collection is obvious as it allows large volumes of journal articles to be easily transported while traveling. Data format incompatibilities between different computer operating systems or software types are fortunately largely resolved by cross-platform formats such as the *Portable Document Format* (PDF) for electronic books, reports and journal articles. Similarly, the standard formats for digital images such as the JPEG or GIF allow images to be exchanged between computers via the Internet or storage devices, and the establishment of large image galleries on the Internet (Chapter 7).

The amount of geoscientific information available on the Internet is enormous. Web search engines such as the market leader *Google Search* (http://google.com) or the less popular *Yahoo!* (http://yahoo.com) and Microsoft's *Bing* (http://bing.com) rank webpages by their relevance using patented algorithms and can therefore respond quite efficiently to queries by Internet users. More specific catalogs exist for searching, evaluating and accessing literature such as the free *Google Books* (http://books.google.com) and *Google Scholar* (http://scholar.google.com), and commercial products such as Elsevier's *Scopus* (http://scopus.com) or Thomson Reuters *ISI Web of Science* (http://isiknowledge.com) (Chapter 2). Geoscientific data are available from large data servers such as the US *National Climate Data Center* (NCDC) (http://ncdc.noaa.go) or the German *Pangaea* (http://pangaea.de) (Chapter 3).

Project Management

In this section, we introduce some useful tools for organizing projects with many participants, for controlling the workflow of the project and for communicating online within the project. We also demonstrate the use of tools to manage the geoscientific information that is collected during the course of the project.

Google Groups, in combination with Google Docs, provides a useful platform for communication within a project, for collecting and ranking information and for processing documents such as project reports, books, or journal articles. During the last couple of years, *Facebook* (http://facebook.com) has also become increasingly popular for presenting research projects on the web. It also encourages research teams to meet, and to exchange or to share information in restricted areas. In addition to these two popular platforms, numerous other networks exist that are mostly used by specific disciplines rather than being widely used Internet tools for project communication in general.

To coordinate project meetings, *Doodle* (http://doodle.com) has become very popular since it started up in 2007. The project leader instigates a Doodle poll by suggesting possible time slots for the meeting and inviting all project members to indicate those dates on which they are available. The project leader then closes the poll and specifies the date selected for the meeting on the basis of mutual availability. Time management during the course of the project is also of great importance and can often be a source of conflict. It is important for the project leader to adhere strictly to the schedule in order to avoid the frustration of individual project members that rely on each other's results, and to avoid any extra costs due to delays. Numerous free calendar and time management software tools, partly cloud-based, are available, as well as commercial alternatives, but will not be discussed here in further detail.

The financial management of a project is typically the responsibility of administrative staff at universities and research institutes. However, the project leader also needs to keep track of the funds that have been spent because of the inherent delay between ordering materials and payment of invoices. Most of the spreadsheet software listed in Section 1.3 can be used for managing finances but more advanced accounting software (not discussed here) may also be useful for managing larger projects.

During the course of the project, all kinds of data accumulate (both original and processed) and also need to be managed. Data management in larger projects is a challenging task that requires the use of database software tools rather than a simple system of files and folders on a hard drive. In this context, a reliable backup system also needs to be established comprising backup hard disks, such as a *Redundant Array of Independent Disks* (RAID) that tolerates the complete failure of one out of several disks, and backup software such as the Apple's *Time Machine* software (http://apple.com/macosx/apps/#timemachine). Storing a backup hard disk in a separate place from the place where the data are processed is advisable. Collaborating on data and documents requires the strict use of version numbers (see also Section 3.9) or the application of the online tools provided by Google Docs, which allow documents to be edited concurrently by multiple users.

1.3   Methods for Processing Scientific Information

Geoscientific information, whether generated within the project or derived from other resources, can be digitally processed by many different kinds of computer software tools. These tools include software for transferring the actual information from one computer to another, from a storage device to computers, or between computers and remote servers in a larger network. The data are processed either locally on a desktop or laptop computer, or remotely on a server using a text terminal or console, *graphical user interface* (GUI) based tool or web-based tool. The processed data are then included in posters, presentations, or papers and other publications.

Software for Transferring Scientific Information

Since the establishment of computer networks, ever increasing quantities of data are transferred from computer to computer via the Internet. Before the introduction of the *World Wide Web* (WWW) and the first web browsers, digital information was mainly accessed through a network by the entry of commands via a keyboard in a text terminal or console (see also Section 3.4). Examples of text terminal applications are *Terminal* and *Console* for Unix-based operating systems such as Solaris, Linux, or Mac OS X, and *cmd.exe* or *PowerShell* for Microsoft Windows.

Whereas in the early 1990s we used these text terminals to send emails and to download data, most scientists today use GUI-based tools such as email clients, FTP software and web browsers to access and share digital information over the Internet. Popular email clients include the open source cross-platform Mozilla *Thunderbird* (http://mozilla.org/projects/thunderbird) and Qualcomm *Eudora* (http://eudora.com), the Microsoft Windows *Live Mail* which is the successor to the popular *Outlook Express* and Windows *Mail* clients (http://explore.live.com/windows-live-mail), and the Apple *Mail* included in Apple's Mac OS X operating system (http://apple.com/macosx).

Data transfer using the *File Transfer Protocol* (FTP) is usually achieved through GUI-based programs such as the popular *Cyberduck* available for Windows and Mac (http://cyberduck.ch), the free *FileZilla* (http://filezilla-project.org), or the commercial *WS_FTP* (http://ipswitchft.com), but numerous other free and commercial alternatives also exist (see also Section 3.4). In many cases, however, the user is not aware of the change in protocols while browsing the Internet to search for geoscientific information online. Since the introduction of the first web browser, *NSCA Mosaic*, in 1993, which later became the *Mozilla* suite of web tools, the most popular web browsers (based on a list of market shares by *NetMarketShare*) are the Microsoft *Internet Explorer* (http://windows.microsoft.com/de-DE/Internet-explorer/products/ie/home) included in Microsoft Windows (but discontinued for Mac and Unix) with more than half of the market share, the cross-platform *Firefox* (http://mozilla.org) with about one fifth of the market share, the cross-platform *Chrome* by Google (http://google.com/chrome), and *Safari* by Apple for Mac and PC, which had about one tenth of the 2011 market share (http://apple.com/safari).

Software for Processing Scientific Information

By far the most popular software for processing data is the *Microsoft Excel* spreadsheet software (http://office.microsoft.com/excel). Following its first introduction for Macs in the mid-80s, and later for Microsoft Windows, the software has seen multiple modifications and improvements before being included in the *Microsoft Office* suite, which also contains *Microsoft Word* and *Microsoft PowerPoint* (http://office.microsoft.com). In 2007, Apple introduced *Numbers* as an alternative to Excel, and included it in the Apple *iWork* suite (http://apple.com/iwork). Numbers is compatible with Excel, but more stable and faster on Macs. A free open-source alternative to Excel is *Calc* included in the cross-platform *OpenOffice* suite that also contains *Writer* and *Impress* (http://openoffice.org). The more advanced, but also more expensive alternative to Excel is the popular *SPSS* software package that has recently been acquired by IBM (http://www-01.ibm.com/software/analytics/spss).

Spreadsheet software packages are, however, generally very limited in their ability to process large data sets, such as those for digital elevation models or images. For such data sets, various high-level programming languages integrated into software environments, either GUI-based or solely command-line based (or both), are very popular tools for processing scientific information. The programming language *R*, which was first introduced in 1993, is probably the most popular member of this family of tools (http://r-project.org), and is also free. R has a command-line interface although several GUIs are available; it has a large number of available libraries and very good documentation. The cross-platform *Python* (http://python.org), first introduced in 1991, is also free and command-line based, but with graphics output using *Matplotlib Graphics* (http://matplotlib.sourceforge.net). The *Interactive Data Language* (IDL) is particularly popular in physics, including astronomy and medical imaging; it was introduced in 1968 and is now sold by *ITTVIS* (http://ittvis.com).

According to the webpage of The MathWorks Inc., the commercial *MATLAB* software is a high-level technical computing language and interactive environment for algorithm development, data visualization, data analysis, and numeric computation (http://mathworks.com). The original software goes back to the late 1970s when Cleve Moler, co-founder of The MathWorks Inc., wrote a program for performing numerical linear algebra. MATLAB can be run either command-line based or using a graphical user interface. Since the cross-platform MATLAB software contains a large library of ready-to-use routines for a wide range of applications, the user can solve technical computing problems much more quickly than with traditional programming languages such as C++ and FORTRAN. The standard library of functions can also be significantly expanded by add-on toolboxes, which are collections of functions for special purposes such as image processing, creating map displays, performing geospatial data analysis or solving partial differential equations. MATLAB is relatively inexpensive for students, either as the MATLAB Student Version (USD $99.00 for MATLAB and USD $29.00 for each toolbox in 2011) which can be ordered by individual students, or as a Student Group License for multiple users, which can be ordered by course instructors thus further reducing the cost per license. The cross-platform *Octave* language is very similar to MATLAB and most routines can be easily ported between the two (http://octave.org).

MATLAB was primarily designed for numerical computations. The *Symbolic Math Toolbox* introduced in 1993 provides tools for solving and manipulating symbolic math expressions and performing variable-precision arithmetic. The classic alternative for such tasks, however, has always been the cross-platform commercial *Mathematica* software developed in 1988 by Stephen Wolfram, who was also the co-founder of Wolfram Research (http://wolfram.com). Over the last two decades, however, Mathematica has also become a powerful tool for high-performance numerical computing similar to MATLAB but with a completely different syntax.

This book does not deal directly with geographical information systems (GIS) or remote sensing. We will, however, use and process digital elevation models and create maps in Chapter 6 using MATLAB. We will also process and georeference a satellite image in Chapter 7, again using MATLAB. More advanced applications in GIS or remote sensing may, however, require software specifically designed for those particular applications. The market leader for GIS is the software development company ESRI (http://esri.com) that makes the commercial ESRI product suite, which includes *ArcGIS*. The free alternative for GIS applications is *GRASS*, which is used for geospatial data management and analysis, image processing, map production, spatial modeling, and visualization (http://grass.osgeo.org). The most popular types of commercial remote sensing software are *ERDAS IMAGINE* and *ERMapper* (http://erdas.com) and ITT Visual Information Solutions *ENVI* (http://ittvis.com).

I hope this brief overview of the most popular software packages will help the reader to choose the most suitable software for his/her applications. My personal preference is for MATLAB because this software makes it relatively easy to get started with sophisticated tools of numerical computing and to create attractive and publishable graphics. I cannot say much about Python but I understand from my colleagues in physics that it is also very popular in certain fields. Octave is certainly an attractive free alternative to MATLAB but has the disadvantage that it lacks the excellent support provided by The MathWorks Inc. Coming from FORTRAN, I did not enjoy the cryptic syntax of Mathematica but nevertheless used it until the mid 90s, for example to invert filters using a Taylor series before then using MATLAB for the actual filtering.

Software for Editing Raster/Pixel Graphics

The *Photoshop* software, part of the *Adobe Creative Suite*, is probably the most widely used professional software for editing raster/pixel graphics, mostly photos (http://adobe.com/photoshop). Originally developed by Thomas Knoll, the software was purchased by Adobe in the late 1980s and first released for Macs in 1990. This software provides an industry-standard editing, enhancement, and output tool for photos. Some of the highlights of the software include excellent selection capabilities, content-aware fill options, processing tools for raw images, automated lens corrections, free-form transformations, advanced painting technologies, and the ability to create 3D artwork from selected graphics objects with direct control over lighting, materials and meshes.

Compared to this huge software package, the very popular and inexpensive *GraphicConverter* by *Lemkesoft GmbH* is a relatively spartan and light product (http://lemkesoft.de). According to the company's webpage, the founder of the company, Thorsten Lemke, changed from Atari to Macintosh in 1992 and wanted to convert his picture collection to a Mac format. Exclusively available for Macs, the software has 1.5 million users worldwide.

In contrast to these commercial products, *Gimp* is a free multi-platform image processing software whose graphical user interface looks similar to that of Photoshop (http://gimp.org). Gimp stands for GNU Image Manipulation Program, which can be used for painting, photo retouching, batch processing, and as an image format converter. The original version of the software was developed by Peter Mattis and Spencer Kimball and it was first released in 1996. Since Photoshop and GraphicConverter do not exist for Linux, Gimp is the most popular software on computers running under this operating system despite being much slower and less convenient when working with photos.

Software for Editing Vector Graphics

The *Illustrator* software, which is also part of the *Adobe Creative Suite*, is the most advanced (but expensive) software for editing vector graphics (http://adobe.com/illustrator). The software was developed by Adobe in 1986 as a commercial version of the company's font development software using the PostScript file format (see Section 3.3) developed by John Warnock at

PARC (the Xerox research institute) with support from Aldus (now a part of Adobe Systems) and Apple Computers (now Apple Inc.). Apart from all the standard vector graphics tools, the software allows perspective drawing, variable-width strokes, stretch controls for brushes, bristle brush, crisp graphics for web and mobile devices and many other advanced tools for vector graphics.

In contrast to Illustrator and other commercial products, the cross-platform *Inkscape* software is free (http://inkscape.org). This open source software was released in 2003 by Ted Gould, Bryce Harrington, Nathan Hurst and MenTaLguY, all members of the former SodiPodi project, which was also an open source vector graphics editor, discontinued in 2004. According to the software webpage, it has capabilities similar to those of Illustrator, using the W3C standard *Scalable Vector Graphics* (SVG) file format. Having used Inkscape while teaching the course on data processing, I personally found that, while it had most of the functionality of the commercial Illustrator, it was far less stable and much slower, as well as being significantly less intuitive and practical. Inkscape, however, seems to be the state-of-the-art vector graphics software if you are using Linux as the operating system of your computer.

Software for Creating Presentations

The *PowerPoint* software, part of the *Microsoft Office* suite, is the market leader in presentation software and is available for both Microsoft Windows and Mac OS X (http://office.microsoft.com/powerpoint). It was created by Thomas Rudkin and Dennis Austin at Forethought Inc. and first released for Macs under the name Presenter. Renamed as *PowerPoint*, it was purchased by Microsoft in 1987 and later included in the Microsoft Office suite. The term *PowerPoint presentation* is now used for any general presentation consisting of a number of slides presented by a projector, even though numerous alternatives, both commercial and free, are now also available. The slides may typically contain a mixture of text, tables, graphics, sound, or videos. The objects of a slide can be arranged using a simple vector graphics editor. This easy-to-use vector graphics editor has made PowerPoint a popular and relatively inexpensive way to create graphics, and even posters, if Illustrator and similar products are not available. Critics of the software argue that the introduction of bullet point lists with PowerPoint encourages less conscientious presenters to oversimplify complex facts into a short list of statements and to rely on recurring, often boring, template designs included in the software package for their presentations (see *Death by PowerPoint*, a term first used by Angela R. Garber at Small Business Computing, http://smallbusinesscomputing.com).

As far as better template designs are concerned, the *Keynote* software included in the Apple *iWork* suite (http://apple.com/iwork) aims to provide more attractive presentation templates (called themes) to the software user. Moreover, Keynote includes a much larger collection of animations within or between slides than does PowerPoint, although in science presentations these animations should be kept to a minimum and used only when appropriate for improving the presentation of scientific results. Keynote can import and export PowerPoint presentations and many Mac users therefore work with Keynote and then convert their presentations for conference laptops.

Impress, which is included in the *OpenOffice* suite, is a free alternative to these commercial products. Formerly known as StarOffice, OpenOffice was developed by StarDivision in the mid 1980s but later acquired by Sun Microsystems in 2002 and then by Oracle in 2010; it is now owned by the Apache Software Foundation. Impress essentially copies PowerPoint, with full compatibility. Since it was available for Unix-based systems at a very early stage, the software is very popular for Linux computers.

Software for Text Processing

Text processing is certainly one of the oldest applications for computers and it is therefore not surprising that a large number of software tools are available for editing text. The most popular text processor is *Word*, as part of the *Microsoft Office* suite (http://office.microsoft.com/word). The first version was created by Charles Simonyi in the early 1980s for Microsoft DOS, and subsequently released for Macs in 1985 and for Microsoft Windows in 1987. In 2005 Apple introduced *Pages* as an alternative to Word and this was later included in the Apple *iWork* suite (http://apple.com/iwork). Pages is compatible with Word but more stable and faster on Macs. The *Writer* software included in the *OpenOffice* suite is a free alternative to these commercial products. Low-level text processors included in the respective operating systems are the Apple *Text Edit* and Microsoft *WordPad*, and there is also *Text Wrangler* (http://barebones.com/products/textwrangler) or *XEmacs* (http://xemacs.org), which are free text editors for use with programing tools. *MathType* by *Design Science* (http://dessci.com) is a software for creating mathematical notation for word processors.

Software for Desktop Publishing

Desktop publishing describes *WYSIWYG* (an acronym for *what you see is what you get*) types of document creation for publication including text, tables, graphics and images. Even though most advanced text processors, such as Microsoft Word and Apple Pages, enable the user to create complex documents that include various types of text and graphics objects, large scale publishing of books, magazines, theses and other multi-page manuscripts requires the use of even more advanced software. In the mid 1980s Macs were the state-of-the-art desktop publishing instruments, especially following the introduction of the *PostScript* standard, the first Apple *LaserWriter*, and software products such as the Aldus *PageMaker*.

Today, the successor product to PageMaker is *InDesign*, which is part of the *Adobe Creative Suite* and offers the most advanced (but expensive) software for desktop publishing (http://adobe.com/indesign). In fact this book has been created by the designer and co-author, Elisabeth Sillmann, using this software. *Scribus* is a free desktop publishing software that is similar to InDesign but is not as convenient, stable, or fast as the commercial product (http://scribus.net). Scribus, however, is also available for Linux-based computers and therefore very popular in the Linux community.

Software for Managing Electronic Libraries on Computers

In order to organize a collection of references and electronic documents, we need a literature or bibliographic management software. A large number of open source and commercial software products are available for this purpose. The most popular commercial software products for literature management are *EndNote* (http://endnote.com) and *Papers* (http://mekentosj.com/papers). Alternatively, software based on the *BibTeX* standard such as *BibDesk* (http://bibdesk.sourceforge.net) for the Unix-based Mac OS X or *KBibTeX* for Linux or Unix uses *LaTeX* to prepare formatted literature lists (http://home.gna.org/kbibtex).

1.4   Presenting Geoscientific Information

The results of a project are typically presented in three formats: on posters, as talks, and in papers. In addition to these principal formats, various other more specific formats exist such as reports, theses, and books as additional examples of scientific information in a text format, keynote lectures as examples of more extended oral presentations, and web presentations including audio or video available online.

Young scientists at the beginning of their scientific careers usually present the first results of their research project as a *poster* at a workshop or conference. A poster is collection of graphs, photos, and text printed on a large piece of paper that is presented on a poster board, usually in a large hall within a conference building. During poster sessions, the presenter of the poster has the opportunity to personally interact with people attending the poster session and visiting the poster board. Prior to the actual presentation of a poster, the scientist submits an *abstract* summarizing the key findings of the research project to date. Abstracts are typically limited to 100 to 200 words, although exceptions are sometimes made, e.g., with extended abstracts which can in some cases reach the length of journal articles. These brief summaries are included in conference volumes or catalogs in which all posters and oral presentations are listed. Abstract volumes are these days no longer printed, but the abstracts are provided on memory sticks or compact disks. Abstracts are viewed by conference session conveners or organizers and checked for relevance to the theme of the session and for their overall quality. In contrast to journal articles, however, abstracts are not peer reviewed. Chapter 10 demonstrates the design of a conference poster and Chapter 11 explains how to write an abstract.

Abstracts are also required to be submitted for an *oral presentation* or *talk*. Such an oral presentation typically lasts 15 to 20 minutes and is supported by a series of pages or slides projected using a video projector. The slides contain graphs and photos, but a relatively small amount of text. The talks are organized in theme sessions that are chaired by a session convener. The convener introduces the presenter, takes care of the time management for the session, and moderates the discussion with questions and answers after the presentation. Chapter 9 explains the planning of a presentation and the design of slides, as well as making suggestions for practicing and delivering the presentation.

After presenting the research results at conferences, either as a poster or as a talk, the research is published as a *paper*, report, thesis, or book. A paper is a journal article that is usually peer reviewed by colleagues who are experts in the same field of research but are not associated with the particular project. Papers are written text, generally four to ten printed pages long including graphs, photos and tables. They are submitted as Word or PDF files with figures, photos and tables included at the end of the file or attached as a separate file, mostly via the web interface of the journal's website. The manuscript arrives in the journal's editorial office where the editor first views and evaluates the paper for its overall quality, length, and suitability for the particular journal. If the paper passes this process, it is sent to two or three independent reviewers from the same field of research who are qualified to evaluate the work. The reviewers, who in most cases remain anonymous, are requested to complete their review within two to four weeks and submit a recommendation to accept the manuscript as it is, to resubmit the manuscript with minor, moderate, or major revisions, or to reject the manuscript either with or without the possibility of resubmission after performing additional experiments and/or a complete rewrite of the text. The editor then collects the reviews, writes an editorial letter summarizing the assessments by the reviewers, and decides on the acceptance or rejection of the work. If it is accepted, the manuscript is then transferred to the copy editor's office where it is prepared for publication. Chapter 11 explains how papers are structured, written and prepared for submission to an international journal for review.

The author of an abstract or paper includes the published work in his/her list of publications, which is a record of the researcher's scientific output. The list of publications is included in the curriculum vitae of a scientist and, together with a list of funded research projects and a record of teaching experience, is an important prerequisite when applying for a research position at a research institute or university.

Recommended Reading

Duarte N (2009) slide:ology: The Art and Science of Creating Great Presentations. O'Reilly Media, Sebastopol, California

Hoffmann AH (2010) Scientific Writing and Communication: Papers, Proposals, and Presentations. Springer, Berlin Heidelberg New York

Reynolds G (2011) Presentation Zen: Simple Ideas on Presentation Design and Delivery. (Voices That Matter)–2nd Edition. New Riders Press, Upper Saddle River, New Jersey

Rhodes JP, Gargett A, Abbott M (2005) Scientifically Speaking. The Oceanography Society, http://tos.org/resources/publications/sci_speaking.html

Trauth MH (2010) MATLAB Recipes for Earth Sciences – Third Edition. Springer, Berlin Heidelberg New York

2   Searching and Reviewing Scientific Literature

2.1   Introduction

This chapter is on searching relevant literature, reviewing and ranking published books and journal articles, and extracting relevant information in the form of text, data, or graphs. In this context, the focus of our book is on Internet resources and literature in an electronic format such as the Adobe *Portable Document Format* (PDF), rather than on printed journals and books in a library. Although some of you might have a printed version of this book in your hands, most people have probably taken advantage of Springer's eBook Collection to read the digital version on a computer.

The advantages of electronic resources are very obvious, as drawers of suspension files filled with hundreds or thousands of printouts of journal articles, or office shelves containing large numbers of heavy books are of little use when working at home or traveling to conferences. This chapter focuses on the most popular open source and commercial Internet resources. The reference search described in the following section demonstrates a typical literature survey by students, starting with the popular online encyclopedias and bookstores, continuing to the generic web search engines to find relevant literature, and ending up in commercial literature data bases designed for professional use.

2.2   Resources for Literature Reviews

In this section, we go through a typical information and literature search, starting with the very popular web-based, open source encyclopedia *Wikipedia* (http://www.wikipedia.org) to find information on earth science topics. Wikipedia was launched in 2001 by Jimmy Wales and Larry Sanger. It attracted 78 million visitors in 2011 and has 91,000 contributors working on about 17 million articles, according its own article about the site. It is written and edited by largely anonymous Internet volunteers, under the supervision of experienced editors who ensure that any edits are cumulative

M. H. Trauth and E. Sillmann, *MATLAB® and Design Recipes for Earth Sciences*,
DOI: 10.1007/978-3-642-32544-1_2, © Springer-Verlag Berlin Heidelberg 2013

improvements. Although often criticized for most articles not being edited by experts in the particular topic, the obvious advantage of Wikipedia is the topicality of its content.

Searching for literature, especially when trying to familiarize oneself with a completely new topic, usually starts in a bookstore. We have agreed to use Internet resources in this book and hence the bookstore will be online. The largest online retailer for books, and more recently for many other products including music, consumer electronics, and even food, is the online commerce company *Amazon* (http://www.amazon.com). Amazon was founded in 1995 by Jeff Bezos. In 2007 Amazon launched the Amazon Kindle, an ebook reader that can download content from the web via a wireless network.

Google was set up in 1996 by Larry Page and Sergey Brin as a graduate student research project at Stanford University. In addition to the general web search services, Google also hosts *Google Books* (http://books.google.com), which was introduced in 2004 under its original name of Google Print. This service searches the full text of both original ebooks and scanned books, as well as excerpts from commercial books, provided as previews by publishers. The sister webpage is *Google Scholar* (http://scholar.google.com), which was introduced in 2004 and searches online journals rather than books. The idea behind this project was to provide a freely-available article search engine as an alternative to some of the commercial products such as Elsevier's *Scopus* or Thomson Reuters' *ISI Web of Science*.

The Web of Science was founded by Eugene Garfield in 1960 and subsequently acquired by the Thomson Reuters Corporation, an information company headquartered in New York City. The Web of Science can be accessed by universities and research institutions from the *Web of Knowledge* (http://isiknowledge.com) index. This index includes about 23,000 scientific journals and includes the Web of Science journal listing and citation index, as well as 110,000 conference proceedings and many other academic resources.

A commercial bibliographic data base specializing in earth science literature is *GeoRef* (http://www.agiweb.org/georef) produced by the American Geological Institute (AGI). GeoRef was launched in 1966 as a series of printed catalogues, later evolving into a CD-based system and ending up as an online service. According to the AGI webpage, this data base contains over 3.2 million references including journal articles, books, maps, conference papers, reports, and theses.

2.3   Finding the Relevant Literature

We now use the Internet resources introduced in the previous section to demonstrate a typical search for books and journal articles, using various indices to rank the literature. We first start with commercial online bookstores, proceed to Google Books and Google Scholar, and end up at professional literature search data bases such as the ISI Web of Science.

Searching with Online Bookstores

In our first example, we would like to explore a new field for future research activities, e.g., human evolution. Let us assume that we are planning to spend €200 on relevant books and start with a literature search on *Amazon* (http://amazon.com) or any other online bookstore. After entering *human evolution* in the search field in early 2011, we obtained a list of more than 40,000 books, DVDs and other products, showing the results in order of relevance. A search for *human evolution* in early 2011 produced the following list of books (Fig. 2.1):

1. Palmer D (2010) Origins: Human Evolution Revealed. Mitchell Beazley

2. Stringer C, Andrews P (2005) The Complete World of Human Evolution. Thames & Hudson

3. Zimmer C (2007) Smithsonian Intimate Guide to Human Origins. Harper Paperbacks

4. Gibbons A (2007) The First Human: The Race to Discover Our Earliest Ancestors

5. Sawyer GJ, Deak V, Sarmiento E, Milner R (2007) The Last Human: A Guide to Twenty-Two Species of Extinct Humans

6. Wood BA (2011) Human Evolution (A Brief Insight). Sterling

7. Shubin N (2009) Your Inner Fish: A Journey into the 3.5-Billion-Year History of the Human Body. Vintage

8. Lewin R (2004) Human Evolution: An Illustrated Introduction. Wiley-Blackwell

9. Cochran G, Harpending H (2010) The 10,000 Year Explosion: How Civilization Accelerated Human Evolution

10. Foley RA, Lewin R (2004) Principles of Human Evolution. Wiley-Blackwell

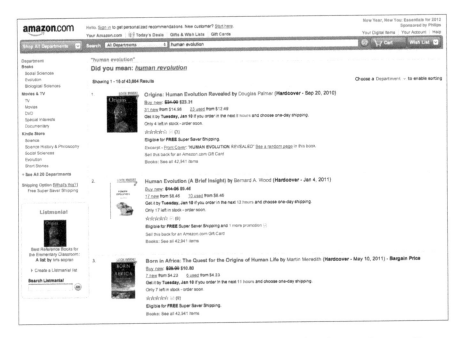

Fig. 2.1 Screenshot from Amazon (http://amazon.com) for the search term *Human Evolution*. The date of screenshot is 9th January 2012, whereas the search experiment described in the text was performed in early 2011, which explains the slight difference in search results.

The search technology is provided by *A9* (http://a9.com), which is owned by Amazon. Amazon also ranks all products by their sales from their own webpages (the *Amazon sales rank*, or ASR, which is updated every hour). Sales, not citations, make a book popular at Amazon: both professionals working in the field of human evolution and interested lay people contribute to the sales of a book, and therefore to the ranking of a particular product.

What makes a customer decide to buy a particular book through Amazon? Why are certain books popular at Amazon and others not? There is much speculation about the exact mathematical algorithm used by the company to calculate the ranking of a book. The obvious factors that may increase the sales of a book are a low price, recent publication date, good title, attractive cover, great illustrations, and positive customer reviews. The reputation of the author is certainly an important criterion within the professional community but might be a secondary consideration for other customers. Douglas Palmer, author of the first-ranked book on human evolution and a very successful author of popular books on paleontology and the earth's history, has not published many original research articles. Ann Gibbons, author of the fourth-ranked book, is a correspondent for *Science* magazine, for which she has covered human evolution for more than a decade. She is certainly in close contact with scientists conducting original research in the field of paleoanthropology and paleontology.

In contrast, Bernard Wood, who authored the sixth-ranked book entitled *Human Evolution (A Brief Insight)* is an anthropologist and the University Professor of Human Origins at the George Washington University. His paperback book of 176 pages, which provides a very brief summary of current knowledge concerning human origins, received mixed but predominantly positive reviews. Robert Foley and Roger Lewin, who authored the tenth-ranked book, are also active researchers in the field of human evolution and have authored numerous highly-cited journal articles and books. Their book, *Principles of Human Evolution*, states very clearly, that it is a textbook for students and professionals rather than a popular science book for an interested lay audience. Robert Lewin also authored the eighth ranked book, *Human Evolution: An illustrated Introduction*, which is a popular book written by an active researcher in the field of paleoanthropology.

These four examples illustrate that authors of science books can range from science writers and journalists with a science background, often within the field of the book's topic, to active researchers writing both popular science books for an educated lay audience and textbooks for students or professionals with a particular interest in the topic. A very good example of a popular book written by an expert is the book

```
Potts R, Sloan C (2010) What Does It Mean to be Human? National
Geographic
```

which was ranked *165* by Amazon in early 2011. Rick Potts is the director of the Smithsonian Institution's Human Origins Program (http:// humanorigins.si.edu/), and Chris Sloan is the National Geographic's paleoanthropology expert (http://www.nationalgeographic.com). In this example, the collaboration between an active researcher and a popular science writer has resulted in an excellent book for both experts and lay persons.

These five books together come to less than €200 and cover the full range of book types, from a textbook for students to a popular book for the interested lay person. As demonstrated in this section, however, it may require a lot of background information on the various authors before the best books on a specific topic can be selected. The next subsections provide some more objective criteria with which to rank books and journal articles.

Searching with Google Books and Google Scholar

What alternative rankings are available that are not based on commercial factors and undisclosed mathematical algorithms? Let us try *Google Books*, as a non-commercially biased search service. By once again entering *human evolution* in the search field, we obtained a list of about 2.2 million results in early 2011 sorted by relevance. Here are the top ten items on the Google Books list:

```
1. Wolpoff MH (1995) Human Evolution. McGraw-Hill

2. Stringer C, Andrews P (2005) The Complete World of Human
Evolution. Thames & Hudson

3. Bradshaw JL (1998) Human evolution: a neuropsychological
perspective. Psychology Press

4. Lewin R (2005) Human Evolution: An Illustrated Introduction.
Wiley-Blackwell

5. Regal B (2004) Human evolution: a guide to the debates.
Controversies in Science, ABC-CLIO

6. Campbell BG (1999) Human Evolution: An Introduction to Mans
Adaptations - 4th Edition. Aldine Transaction

7. Maxwell M (1984) Human evolution: a philosophical
anthropology. Columbia University Press

8. Wood B (2005) Human evolution: a very short introduction.
Oxford University Press

9. Cela-Conde CJ, Ayala FJ (2007) Human Evolution: trails from
the past. Oxford University Press

10. Khanna DR (2004) Human Evolution. Discovery Publishing House
```

This certainly is a significantly different type of list, including not only books but also book chapters; it is computer generated by an automated search for keywords without the involvement of any quality criteria. The books on the Amazon list by Roger Lewin, Bernard Wood, and Chris Stringer with Peter Andrews, also appear on the Google Books list, but at least in one case as an older edition with a different publisher. The other items listed are relatively less popular books or monographs on specific aspects, current debates, or perspectives of human evolution. This example demonstrates that, although biased by non-scientific criteria, searching for books on the basis of sales-based rankings seems to be the better option for finding the relevant literature.

Finding relevant journal articles on a specific topic is a much simpler task by far. In this case, the rankings are based on citations by experts rather than on sales to a mixed professional and non-professional clientele. As an example, we aim to find scientific literature on the influence of the Younger Dryas cold reversal in Africa. The online encyclopedia provides some general information on the Younger Dryas as a climate event:

```
http://en.wikipedia.org/wiki/Younger_Dryas
```

Authored by anonymous and possibly non-expert contributors, this article states:

```
The Younger Dryas stadial, also referred to as the Big Freeze,[1]
was a geologically brief (1,300 ± 70 years) cold climate period
between approximately 12,800 and 11,500 years ago (between 10,800
and 9500 BC).[2]
```

citing two journal articles [1] and [2], namely

```
Berger, WH (1990) The Younger Dryas cold spell — a quest
for causes. Global and Planetary Change 3 (3):219-237.
doi:10.1016/0921-8181(90)90018-8.

Muscheler R, et al. (2008) Tree rings and ice cores reveal 14C
calibration uncertainties during the Younger Dryas. Nature
Geoscience 1:263-267. doi:10.1038/ngeo128.
```

W.H. Berger is the author of the first article, which was published in the year 1990, entitled *The Younger Dryas cold spell – a quest for the causes.* The article appeared in Elsevier's *Global and Planetary Change* journal (Volume 3, Issue 3, pages 219–237). DOI stands for *Digital Object Identifier*, which is a character string that enables all kinds of electronic documents and objects to be identified. The second article, authored by R. Muscheler and his co-authors B. Kromer, S. Björck, A. Svensson, M. Friedrich, K.F. Kaiser, and J. Southon, appeared in the *Nature Geoscience* journal in the year 2008 (Volume 1, pages 263–267). Since the article has more than three authors, the co-authors were merged as *et al.*, which stands *et alii*, Latin for *and others*. The exact format, in particular the use of *et al.*, depends on the format required for citations by the particular journal.

We are not yet sure whether the two cited articles are the most relevant references for the Younger Dryas. The articles on Wikipedia are, however, always a good starting point for an in-depth literature research on a specific topic. Knowing that the Younger Dryas, in general, is a time interval between 12,800 and 11,500 years ago that was characterized by a cold climate at least points us in the right direction for our literature review on the topic.

We next use *Google Scholar* for a more detailed search of journal articles (Fig. 2.2). For this we enter *Younger Dryas Africa* in the search field of

```
http://scholar.google.com
```

and get a list of about 7,000 references. According to Google Scholar's webpage, it aims to rank documents the way researchers do, weighing the full text of each document, where it was published, who it was written by, as well as how often and how recently it has been cited in other scholarly literature. Browsing the list of articles retrieved by our search, however, it is not obvious how the articles are ranked, but the number of citations and the occurrence of the search terms in the title certainly seem to have had some influence. The top ten articles on the list are:

```
1. N Roberts, M Taieb, P Barker, B Damnati, M Icole, D Williamson
(1993) Timing of the Younger Dryas event in East Africa from
lake-level changes. Nature 366:146-148 [Citations: 74]
```

Fig. 2.2 Screenshot from the *Advanced Scholar Search* of *Google Scholar* (http://scholar. google.com), taken in early 2012. Google Scholar was introduced in 2004 and included online journals rather than only books.

2. Y Garcin, A Vincens, D Williamson, G Buchet, J Guiot (2007) Abrupt resumption of the African Monsoon at the Younger Dryas--Holocene climatic transition. Quaternary Science Reviews 26:690-704 [Citations: 26]

3. D Rind, D Peteet, W Broecker, A McIntyre, W Ruddiman (1986) The impact of cold North Atlantic sea surface temperatures on climate: implications for the Younger Dryas cooling (11-10 k). Climate Dynamics 1:3-33 [Citations: 126]

4. RB Alley (2000) The Younger Dryas cold interval as viewed from central Greenland, Quaternary Science Reviews 19:213-226 [Citations: 218]

5. P deMenocal, J Ortiz, T Guilderson, J Adkins, M Sarnthein, L Baker, M Yarusinski (2000) Abrupt onset and termination of the African Humid Period::: rapid climate responses to gradual insolation forcing. Quaternary Science Reviews 19:347-361 [Citations: 290]

6. D Verschuren, KR Laird , BF Cumming (2000) Rainfall and drought in equatorial east Africa during the past 1,100 years. Nature 403:410-414 [Citations: 302]

7. P De Deckker, T Corrège, J Head (1991) Late Pleistocene record of cyclic eolian activity from tropical Australia suggesting the Younger Dryas is not an unusual climatic event. Geology 19:602-605 [Citations: 33]

8. MR Talbot, ML Filippi, NB Jensen, JJ Tiercelin (2007) An abrupt change in the African monsoon at the end of the Younger Dryas. Geochemistry Geophysics Geosystems 8:Q03005 [Citations: 12]

9. W Zhou, MJ Head, Z An, P De Deckker, Z Liu, Y Liu, X Lu, D Donahue, AJT Jull, JW Beck (2001) Terrestrial evidence for a spatial structure of tropical-polar interconnections during the Younger Dryas episode. Earth and Planetary Science Letters 191:231-239 [Citations: 26]

10. H Renssen (1997) The global response to Younger Dryas boundary conditions in an AGCM simulation. Climate Dynamics 13:587-599 [Citations: 24]

Depending on our institution's journal subscriptions, we may be able to directly access the PDFs of these papers from the publisher's website by following the link provided in the Google Scholar list of articles. As we will see later, the first article, which is by N. Roberts, was indeed the earliest publication on the influence of the Younger Dryas cold event in Africa. Published in *Nature* in 1993, it was the result of a French-British collaboration in the 1980s and 1990s led by M. Taieb, and is fairly-well cited with 74 citations. The paper by Y. Garcin is one of many follow-up publications from this project, although the main conclusions are different from those in the Roberts paper (see Section 2.4). Articles no. 3, 4, 9 and 10 on the list provide a more global perspective on the influences of the Younger Dryas, including some comments on Africa, while article no. 7 is on climate change in Australia at that time. Article no. 8 provides some insights into the termination of the Younger Dryas in Africa, whereas the highly-cited article no. 5 is on climate change in Africa over the last 23,000 years, including the Younger Dryas. The well-cited paper by D. Verschuren, listed as no. 6, does not address the influence of the Younger Dryas at all as it only provides a history of climate change in Africa during the last 1,100 years. It does, however, refer to the first article on the topic (by N. Roberts) and for that reason appears in the list of relevant literature.

As we can see from this experiment, our Google literature search gives some reasonable results but with limitations. Whereas articles no. 1, 3, 4, 5, 9 and 10 may provide a good overview of the Younger Dryas in general, and of its influence in Africa in particular, articles 2 and 7 represent more specific articles with controversial discussions and conclusions rather than providing established knowledge on climatic connections between high and low latitudes at the last glacial termination. Article no. 6 is not at all relevant as it does not include any discussion on the Younger Dryas in its analysis of past climate change over the last millennium. The article by N. Roberts is cited by D. Verschuren and co-authors for noting the difference in the influence that solar heating had on moisture levels in Africa between glacial and interglacial periods, and the variations in moisture levels during the transition from the Medieval Warm Period to the Little Ice Age. Article no. 8 might be of interest for comparing Africa with other areas in the subtropics and tropics.

Searching with the Thomson Reuters ISI Web of Science

There are not many possibilities for influencing the search results in Google Scholar. The exact routine that ranks the journal articles for their relevance is not obvious. In our next experiment, we use the commercial Thomson Reuters ISI *Web of Science* literature search data base, which is very similar to Elsevier's Scopus service. For this we enter *Younger Dryas Africa* in the *topic* search field of

```
http://isiknowledge.com/
```

In early 2011 this yielded a list of about 100 articles, ranked by publication date (Fig. 2.3). In contrast to Google Scholar, the ISI Web of Science allows changes to be made to the mode of ranking, which can be, e.g., by the first author, the publication date, the number of citations, and the title of the journal. There is also the option to sort by relevance using a ranking system that considers how many of the search terms are found in each record, according to *Help*. Sorting by relevance, the article

```
Abell PI, Plug I. (2000) The Pleistocene/Holocene transition in
South Africa: evidence for the Younger Dryas event. Global and
Planetary Chance 26:173-179
```

which has only been cited 14 times during the last decade, ranked first in early 2011. Sorting the list of 100 articles on the Younger Dryas in Africa by the number of citations, the article

```
BarMatthews M, Ayalon A, Kaufman A (1997) Late quaternary
paleoclimate in the eastern Mediterranean region from stable
isotope analysis of speleothems at Soreq Cave, Israel. Quaternary
Research 47:155-168
```

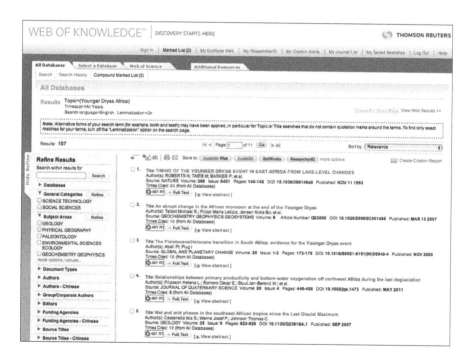

Fig. 2.3 Screenshot from Thomson Reuters' ISI Web of Knowledge (http://isiknowledge. com), showing results of a search for *Younger Dryas Africa*. The date of the screenshot is 9th January 2012, whereas the search experiment described in the text was performed in early 2011, which explains the slight difference in search results.

ranks first, with 185 citations. Sorting by publication date, the article

```
Roberts N, Taieb M, Barker P, Damnati B, Icole M, Williamson D
(1993) Timing of the Younger Dryas event in East Africa from
lake-level changes. Nature 366:146-148
```

which was ranked first in Google Scholar, is the oldest article of relevance on the Younger Dryas in Africa. Clicking on the number *90* opens a new webpage listing all other articles citing *Roberts et al. (1993)*. After having identified this article as one of the earliest to provide evidence of a Younger Dryas influence in Africa, we also find

```
Williamson D, Taieb M, Damnati B, Icole M, Thouveny M (1993)
Equatorial extension of the Younger Dryas Event - Rock Magnetic
Evidence From Lake Magadi (Kenya). Global and Planetary Change
7:235-242
```

which was published in the same year. What other criteria would help us to identify the best article on the topic, rather than relying on simple counts of the number of times the search term *Younger Dryas Africa* is found in a record? In addition to the quality of the content, three measures are available to assist us in finding the best article on the topic; these measures are widely used but their validity is intensely debated.

The first measure for the relevance of an article is the *number of citations*, already mentioned when we used Google Scholar to search for literature. The number of citations counts all citations of an article in journals listed on the ISI Web of Science since the article's publication date. It therefore does not count the citations of an article in books or other media. In this regard the article by N. Roberts is more frequently cited (90 times based on ISI Web of Science, 74 times according to Google Scholar) than the article by D. Williamson (22 times). On the basis of this measure, the article by N. Roberts therefore appears to be more relevant, but let us explore some other criteria before we come to a final decision on the most relevant publication.

The second measure for the relevance of an article is the journal *impact factor* (IF). The journal impact factor can be obtained from the *Journal Citation Reports* (JCR) of the ISI Web of Science. The JCR Science Edition contains 7,300 journals in science and technology; the JCR Social Science Edition, not considered here, contains data from over 2,200 journals in social sciences. The JCR lists the number of articles published in the journal per year, the number of citations of that journal in articles published during the same year, the impact factor calculated from that year, and many other parameters. The IF calculated annually is the average number of citations of an article in journals during the two previous years. The IF measures only citations for journals that are listed on the ISI Web of Science and also only counts citations from those journals.

Since being introduced by Eugene Garfield, the founder of the ISI, the impact factor has been under severe criticism. For instance, the IF is highly dependent on the discipline. The journal *Nature*, in which the article by N. Roberts was published, and its subsidiary journal *Nature Geoscience*, are good examples of this dependency. The British journal *Nature* and its American counterpart *Science*, are the world's most respected scientific journals, with 2009 impact factors of 34.480 (Nature) and 29.747 (Science), i.e., articles in these journals were, on average, cited about 30 times during 2007 and 2008. The impact factor of these and other journals can be accessed from the ISI Web of Science webpage by clicking the *Additional Resources* tab, and then following the *Journal Citation Reports* link. There you can select the JCR edition and year, e.g., the *JCR Science Edition* for the year *2009*, and select the option *Search for a specific journal*. Enter *Nature* in the search field and you will get a table with the rank of the journal based on the IF, the total number of citations, the IF, and the five-year IF, as well as many other parameters for measuring the success of a journal.

Going back to the article by N. Roberts, cited 90 times since its publication date, we can analyze the citation record over the last two decades. Again, clicking on the number *90* (the number of citations) opens a new webpage listing all other articles citing *Roberts et al. (1993)*. There we click on *Analyze Results* and get a new webpage with various options for ranking and sorting the results. We then choose *Publication Year* to rank the records and *Selected field* to sort the results. This yields a chart displaying the citations per year where we can see that the article typically receives 5 to 7 citations per year, with a peak of 14 citations in the year 2000. Note that this citation rate is significantly lower than the journal's two-year average citation rate, i.e., its impact factor. However, Nature's citation rate in 1993 was different from that of today. The article by N. Roberts, which certainly made an important contribution to the understanding of the influence of high-latitude climate change in the tropics and provided the first evidence of the Younger Dryas cold event in East Africa, does not receive a large number of citations simply because it covers a relatively restricted field of research in which a low number of active scientists are involved.

The article's citation rate, however, is within the range of the IF of Nature Geoscience, which was launched as a disciplinary sub-journal of Nature in January 2008. According to the Journal Citation Reports, the 2009 IF for this journal is only 8.108, as compared to 34.480 for Nature. H. Langenberg, editor at Nature Geoscience, explained the difference between the IFs of the two journals as being due to the average citation rates per paper varying by a factor of five or six, depending on the scientific field, with the geosciences at the low end compared to other fields such as genetics and stem-cell research in which articles receive many more citations (Langenberg 2010). Browsing through the Journal Citation Reports for other well-respected journals in earth sciences, the third-best ranked journal behind Nature and Science is the *Proceedings of the National Academy of Sciences of the United States of America* (PNAS). PNAS is another interdisciplinary journal publishing original research; it has a 2009 IF of 9.432, and is therefore ranked slightly higher than Nature Geoscience, which is the most highly ranked specialized journal in our field.

Other highly ranked journals in earth sciences are Reviews of Geophysics (IF 8.021), Annual Reviews of Earth and Planetary Sciences (IF 7.581), and Earth-Science Reviews (IF 6.942). Below these, numerous journals are ranked with an IF in the 3 to 5 range, such as *Geology* (IF 4.368), *Quaternary Science Reviews* (IF 4.245), *Earth and Planetary Science Letters* (IF 4.062) and the more specialized journal *Paleoceanography* (IF 3.644). Almost 90 % of all earth science journals, however, fall within the IF < 3.000 range, presumably including articles that are not cited at all. The article by D. Williamson was published in *Global and Planetary Change*, a journal ranking at the lower end of the top 10 % of journals according to its IF of 3.272. Based on the IF of the two journals in which the articles about the Younger Dryas in Africa were published, the article by N. Roberts is again the better choice.

The third measure, which describes publication success, is the *h-factor* of the first author. This parameter, which is even more intensively debated than the previous two, was introduced by J.E. Hirsch to quantify an individual's scientific research output (Hirsch 2005). The index h is defined as the number of papers published by a scientist with citation number $\geq h$. According to J.E. Hirsch, the index h for a given individual should increase approximately linearly with time, i.e., scientists produce papers of similar quality at a steady rate over the course of their careers. Of course, the slope m of h over n years varies between different researchers, again depending not only on publication success but also on the discipline and on co-authorship regulations. On the basis of the slope m, J.E. Hirsch concludes that $m \approx 1$, i.e., $h=20$ after 20 years, characterizes a successful scientist, $m \approx 2$ characterizes an outstanding scientist, and $m \approx 3$ characterizes truly unique individuals.

To determine the h-factor for the two first authors, N. Roberts and D. Williamson, we click on the *Web of Science* tab on the ISI Web of

Knowledge webpage, and then enter *Roberts N* in the *Author* search field. We enable the *Science Citation Index Expanded* data base and disable the social sciences and arts and humanities citation indices, and then click *Search*. This yields a list of about 550 references, including those for other scientists with the same name as the author of the article on the Younger Dryas. We can *View Distinct Author Sets* by clicking on *Roberts N* above the list to exclude those authors from other fields such as clinical sciences, biology of reproduction, and librarianship, and keep the person author- ing articles in the field of physical sciences. This selection reduces the to- tal number of publications to 33, ranking the Nature article first with 90 citations. According to the Journal Citation Reports, the total number of citations of N. Roberts is 703, the average number of citations per item is 21.30, and the *h*-factor is 15. Since we are not sure whether we have indeed included all publications by N. Roberts, we return to the *Results* page and *Refine Results* by excluding *Subject Areas*. We then get a long list of sub- ject areas, in which we activate the field of physical sciences, which includes geosciences, geography, geochemistry and geophysics, environmental sci- ences, multidisciplinary sciences, limnology, paleontology, oceanography, and archeology, and then click *Refine*.

This then yields a list of 54 publications, but it soon becomes evident that we have again included authors with similar names but from different fields, such as N. Roberts working on environmental trace-metal contamination. Excluding those publications from authors such as Noel Roberts, Nicholas J. Roberts and others, we end up with pretty much the same number of about 35 publications by the C. Neil Roberts that works at the University of Plymouth and led projects on paleoenvironmental change in Turkey after he stopped working in Africa, as we had before. Cross-checking the Journal Citation Reports with the (incomplete) list of publications provided on his webpage (http://www.plymouth.ac.uk/staff/cnroberts), we are very con- fident that we have arrived close to the true publication record of the N. Roberts who authored the article on the Younger Dryas in Africa. As I know from the author, however, both our search results and Neil Roberts' web- page are missing one important article, which was first-authored by Henry Lamb from the University of Aberystwyth and co-authored by N. Roberts:

Lamb, H.F., Gasse, F., Benkaddour, A., el-Hamouti, N., van der Kaars, S., Perkins, W.T., Pearce, N.J., Roberts, C.N. (1995) Relation between century-scale Holocene arid intervals in tropical and temperate zones, Nature, 373, 134-137.

According to the Web of Science, this article has been cited more than 175

2 | SEARCHING AND REVIEWING SCIENTIFIC LITERATURE

times. This article is missing from our search list because in the list of authors Roberts is listed as *C.N. Roberts*. Neil Roberts is known by his middle name Neil, but some publications, such as the one published in Nature, also add his first initial C. Note that the same author has also published the very popular textbook

```
Roberts N (1998) The Holocene: an environmental history - 2nd
edition. Blackwell, Oxford
```

which is not mentioned at all in the ISI Citation Reports. Not considering books is of course a significant weakness of the ISI Web of Science. Books do not contribute to any of the measures of the scientific or publication success of a researcher. Recently, however, Thomson Reuters has launched a *Book Citation Index* for the Web of Knowledge, which can help to find the most relevant books on a topic, using similar parameters to those used for journal articles to measure the contribution that books make a particular discipline.

How does D. Williamson's *h*-factor compare with that of N. Roberts? The first search returns about 670 references, but can *View Distinct Author Sets* reduce the articles to only those published by the French researcher D. Williamson? By excluding all publications related to life sciences we reduce the list to about 43 articles, with a total of 1,016 citations, an average citation rate of 24.19, and an *h*-factor of 20. David Williamson, who led the paleomagnetics laboratory at the CEREGE in Aix-en-Provence, obviously benefitted from being a laboratory manager. He is co-author of many highly-cited paleoclimate publications that used paleomagnetics, but also first-authored several well-cited articles on mineral-magnetic proxies in paleoclimate research.

Please note that he is also a co-author on N. Roberts' Nature paper, to which he probably contributed the magnetics record shown in Figure 2 of the article. Comparing the authors listed for both papers

```
Roberts N, Taieb M, Barker P, Damnati B, Icole M, Williamson D
(1993) Timing of the Younger Dryas event in East Africa from
lake-level changes. Nature 366:146-148

Williamson D, Taieb M, Damnati B, Icole M, Thouveny M (1993)
Equatorial extension of the Younger Dryas Event - Rock Magnetic
Evidence From Lake Magadi (Kenya). Global and Planetary Change
7:235-242
```

reveals that other authors are also listed for both articles. An Internet search for some of these authors leads to the conclusion that both N. Roberts and D. Williamson worked under the guidance of Maurice Taieb, who was the leader of the paleolimnological project at Lake Magadi in Southern Kenya.

M. Taieb is a French geologist and paleoanthropologist who worked in Africa for over fifty years, and was involved in the discovery of the fossil fragments of *Lucy* in the early 1970s.

Because of the leadership and co-authorship of M. Taieb on both publications, and because D. Williamson's article is on the paleomagnetic record only whereas N. Roberts' paper includes a discussion of other climate proxies such as diatoms (silica algae), organic carbon, and calcite, and also because Nature is a far better reference than Global and Planetary Change, we come to the conclusion that the article by N. Roberts is, in general, likely to be the more useful publication on the Younger Dryas in Africa. An analysis of our literature research, however, clearly shows how difficult it is to evaluate the relevance of a specific article on the basis of bibliometric measures such as journal impact factors, citation indices, and *h*-factors. As the demonstration clearly shows, it often requires a lot of background knowledge to find the most relevant papers on a particular topic. Of course having found the most important journal article should not prevent you from also reading some of the less relevant papers. These may well contain interesting details in the text that are more relevant to your own particular research than to that of the citing authors.

2.4 Extracting the Relevant Information from Literature

Having identified N. Roberts' article as the most relevant reference to use as a starting point for research into the influence of the Younger Dryas event in East Africa, we now obtain the electronic version of this article , which was published in Nature (http://www.nature.com) in 1993. The journal's webpage has a search field in which we can enter the terms *Younger Dryas East Africa* and get about 40 references as a result. We see that there are even older articles mentioning the possible influence of the Younger Dryas in Africa based, for example, on analysis of the sapropels in the Mediterranean Sea as indicators of monsoonal rainfall intensity:

```
Rossignol-Strick M, Nesteroff W, Olive P, Vergnaud-Grazzini C
(1982) After the deluge: Mediterranean stagnation and sapropel
formation. Nature 295:105-110
```

The article by N. Roberts also shows up in the 40 references, listed as a *Letter to Nature* type of article. A Letter in Nature is four pages long and has no more than 30 references, whereas an *Article* can be up to five pages long and has up to 50 references; *Brief Communications* are even shorter than Letters.

```
Timing of the Younger Dryas event in East Africa from lake-level
changes
Nell Roberts, Maurice Taieb, Philip Barker, Brahlm Damnati,
Michel Icole, David Williamson
Nature 366, 146-148 (11 November 1993) doi:10.1038/366146a0
Letter
Abstract | PDF | Rights and permissions | Save this link
```

Please note that the author's first name Neil is actually misspelled Nell in the journal's article data base. The person who typed the reference into the data base presumably used the scanned version of the article, which we download later from the journal's webpage, in which the first name indeed looks like Nell rather than Neil.

As an alternative, we can browse the contents of the journal if we know the volume and page numbers for the article. On the main page of Nature, we click on *Publications A–Z* under the *Inside nature.com* header. There we choose the journal *Nature* from the list of publications, then *Archive* from the menu bar of the journal's webpage, select the year *1993*, volume *11 November 1993 366 6451 95–188* and get the contents list for the issue of the journal that contains the article by N. Roberts.

```
Timing of the Younger Dryas event in East Africa from lake-level
changes 146
NELL ROBERTS, MAURICE TAIEB, PHILIP BARKER, BRAHLM DAMNATI,
MICHEL ICOLE & DAVID WILLIAMSON
doi:10.1038/366146a0
First paragraph & References | PDF (355K)
```

The reference contains the title, the page number (*146*), the authors, the DOI (*0.1038/366146a0*), the first paragraph and references, and a link to the full article. The title, list of authors, abstract and references, are freely available without subscription to the journal (Fig. 2.4). The full article, however, is available by subscription only. In many cases, however, the title and the abstract already contain the most important information from the article. We are interested to learn about the influence of the Younger Dryas in East Africa. If the Younger Dryas did have an influence, how did it affect the climate in East Africa and when? The title

```
Timing of the Younger Dryas event in East Africa from lake-level
changes
```

suggests that there was indeed an influence of the Younger Dryas in East Africa, and that this influence has been identified in the water-level record of Lake Magadi in Kenya. Furthermore, the title indicates that the article discusses the timing of the Younger Dryas influence in the region. It does not,

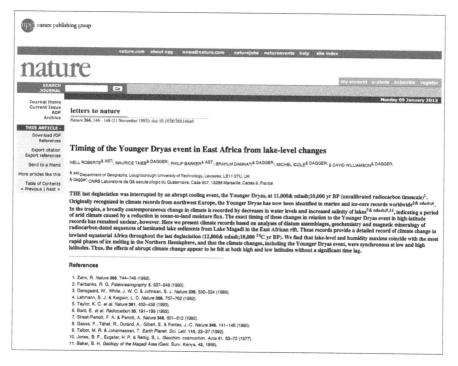

Fig. 2.4 Screenshot from *Nature* (http://nature.com) showing the freely-available title, list of authors, abstract, and references from the paper by Roberts et al. (1993). The date of the screenshot is 9th January 2012 whereas the search experiment described in the text was performed in early 2011, which explains the slight difference in search results.

however, contain information on the nature of the influence (e.g., whether it resulted in a drier or wetter climate) or the exact timing of the resulting climate change in East Africa.

We proceed to the abstract of the article, which typically provides a summary, in most cases without reference to previous articles by other researchers. For articles in Nature, however, the abstract is a fully-referenced first paragraph, in bold format.

```
THE last deglaciation was interrupted by an abrupt cooling
event, the Younger Dryas, at 11,000-10,000 yr BP (uncalibrated
radiocarbon timescale)¹. Originally recognized in climate records
from northwest Europe, the Younger Dryas has now been identified
in marine and ice-core records worldwide²⁻⁶. In the tropics,
a broadly contemporaneous change in climate is recorded by
decreases in water levels and increased salinity of lakes⁷⁻⁹,¹⁴,
indicating a period of arid climate caused by a reduction in
ocean-to-land moisture flux. The exact timing of these changes
```

2 | SEARCHING AND REVIEWING SCIENTIFIC LITERATURE

```
in relation to the Younger Dryas event in high-latitude records
has remained unclear, however. Here we present climate records
based on analyses of diatom assemblages, geochemistry and
magnetic mineralogy of radiocarbon-dated sequences of laminated
lake sediments from Lake Magadi in the East African rift. These
records provide a detailed record of climate change in lowland
equatorial Africa throughout the last deglaciation (12,800-10,000
14C yr BP). We find that lake-level and humidity maxima coincide
with the most rapid phases of ice melting in the Northern
Hemisphere, and that the climate changes, including the Younger
Dryas event, were synchronous at low and high latitudes. Thus,
the effects of abrupt climate change appear to be felt at both
high and low latitudes without a significant time lag.
```

The small superscript letters refer to articles in the list of references. As an example, reference 1 after the first sentence refers to

```
1. Zahn, R. Nature 356, 744-746 (1992).
```

which is the article by R. Zahn

```
Zahn, R (1992) Deep ocean circulation puzzle. Nature 356:744-746
```

on the Younger Dryas event in general. This article is cited after the first sentence of the article by N. Roberts. This sentence introduces the phenomenon of the Younger Dryas as an abrupt cooling event between 11,000 and 10,000 ^{14}C yr BP during the last deglaciation, i.e., during the transition from the last glacial to the present interglacial. The second and third sentences state that the Younger Dryas was a global event that resulted in a drier climate in the tropics. The exact timing of tropical climate change during the Younger Dryas event, however, remains unclear.

Following an outline of the scientific question to be investigated, the second half of an abstract or first paragraph typically starts with the statement *Here we present ...*, introducing the approach, results, and conclusions of the actual research presented in the article. In our example, this reads as follows:

```
Here we present climate records based on analyses of diatom
assemblages, geochemistry and magnetic mineralogy of radiocarbon-
dated sequences of laminated lake sediments from Lake Magadi in
the East African rift.
```

from which the reader learns that the article is about a lake-sediment record from a lake located in the East African rift, and that it includes the analysis of fossil silica algae (diatom) assemblages, as well as chemical and physical sediment parameters. The last three sentences of the abstract

These records provide a detailed record of climate change in
lowland equatorial Africa throughout the last deglaciation
(12,800-10,000 ^{14}C yr BP). We find that lake-level and humidity
maxima coincide with the most rapid phases of ice melting in the
Northern Hemisphere, and that the climate changes, including the
Younger Dryas event, were synchronous at low and high latitudes.

basically list the results – more specifically, the age of the sediments and
therefore the time interval contained in the lake-level record, and the information that the climate change coincided with the high-latitude deglaciation – but it does not contain information on precisely how the Younger
Dryas affected the climate in the region. Remembering that the first part of
the paragraph said that the Younger Dryas was generally dry in the tropics
and reading here that humidity maxima coincided with the ice melting, we
could conclude that the Younger Dryas cold event was probably dry. The
ultimate conclusion of the article is then given in the final sentence of the
abstract

Thus, the effects of abrupt climate change appear to be felt at
both high and low latitudes without a significant time lag.

essentially proposing a link between high and low latitudes based on the
correlation of climate records without any time lag.

Reading the title and the abstract (or the first paragraph) of the article
has taken us only one or two minutes, but we have already extracted the
most important scientific information, despite the minor weakness in this
particular article of not being precise about the exact nature of the influence
that the Younger Dryas had in East Africa. In our experiment this weakness
necessitates an analysis of the full article, or at least parts of it. The article is
in a Portable Document Format (PDF). This format was designed by Adobe
Systems (http://www.adobe.com/) as self-contained cross-platform document. PDF files contain the complete formatting of vector illustrations, raster images and text, or a combination of all these, and includes all necessary
fonts. These files are highly compressed, allowing a fast Internet download.
Adobe Systems provides the Acrobat Reader free-of-charge for all computer
platforms, for reading PDF files. As we will see later, more recent articles
are in a vector format, allowing text fragments and graphics to be extracted
and incorporated into lectures, or into books such as this, provided that the
copyright of the electronic materials is respected. The article by N. Roberts,
published in 1993, is a scanned version of the original printed journal article
and therefore only available in raster format.

The article is approximately 2.5 pages long and therefore represents a
relatively short piece of text. Nature, Science and the Proceedings of the

National Academy of Sciences (PNAS) typically publish articles with a maximum length of four pages, whereas most other journals allow papers of 10 printed pages, or even more. Most journals also allow the authors to publish an electronic supplement to the printed article, including more text, tables, figures and references. Reading an article of such length requires efficient reading, and we therefore need to know the most likely location of the most relevant information in an article. Fortunately, research articles share a common structure comprising the title and abstract, introduction, setting and methods, results, discussion, and conclusions. Differences exist between journals, however, and Nature, Science, and PNAS each structure articles in a slightly different way. The overall structure of the introduction, methods, results, and discussion/conclusions also applies to these journals but the sections do not have headers. Details about the geographic and geologic setting of the study area, an overview of the methods used in the analysis, and results presented as tables and text are usually published separately from the actual article as electronic supplementary information.

Apart from the information contained in the title and abstract/first paragraph, the most important information typically occurs in the conclusion or the last paragraph of the article, and in the last figure. The last figure of an article often summarizes the results or places a new data set in a larger context by, for instance, correlating the new data with established records of past climate change, or with any other data set from a different location and from a different research project. In our example, Figure 3 compares the water-level record from Lake Magadi in Kenya with other climate records, namely the oxygen-isotope curve from Rotsee in Switzerland by Lotter et al. (1989) and winter temperatures for Britain reconstructed from beetle remains by Atkinson et al. (1987):

```
Lotter A, Zbinden H (1989) Late-Glacial pollen analysis, oxygen-
isotope record, and radiocarbon stratigraphy from Rotsee
(Lucerne), Central Swiss Plateau. Ecologae geol. Helv. 82:191-202

Atkinson TC, Briffa K, Coope GR (1987) Seasonal temperatures in
Britain during the past 22,000 years, reconstructed using beetle
remains. Nature 325:587-592.
```

The figure clearly shows that the water level dropped by about 40 meters during the Younger Dryas, suggesting that tropical East African climate became drier when the oxygen isotopes and beetle assemblages indicate colder temperatures in the mid latitudes. The relevant text in the last paragraph, which typically contains the conclusion of the article, also says that

```
Conditions became warmer and wetter around 12,700 14C yr BP,
```

```
cooler and drier at 11,100-10,700 14C yr BP, with renewed warming
and wetting ~10,000 14C yr BP.
```

and then comments on the timing of this dry event in the tropics, relative to the Younger Dryas cold event

```
The similarity of these and other records indicates that this
climate oscillations was not only a global event, but also
that there were no significant time lags between tropical and
temperate components of the climate system during the last
deglaciation.
```

Having read the most important sections of the article, we read the remaining text and figures in more detail if additional information is required for the article to be used as a reference in our own manuscript (see Chapter 12). In that case, we cite the article in the text as

```
Roberts N, Taieb M, Barker P, Damnati B, Icole M, Williamson D
(1993) Timing of the Younger Dryas event in East Africa from
lake-level changes. Nature 366:146-148
```

in the reference list of our article. The format used for references can vary significantly depending on the requirements of the particular journal, provided in their *Guidelines for Authors*. In the text itself, the article is cited as

```
Roberts et al. (1993) have presented ...
```

where, as explained previously, *et al.* stands *et alii*, which is Latin for *and others*, since the article has more than two authors. If there are two authors only, e.g., N. Roberts and M. Taieb, the reference is cited as

```
Roberts and Taieb (1993) have presented ...
```

or

```
Roberts & Taieb (1993) have presented ...
```

In most cases, a reference occurs at the end of a statement rather than being incorporated in the text itself. The references then reads

```
The Younger Dryas in Africa was dry (Roberts et al., 1993).
```

or in the case of two authors

```
The Younger Dryas in Africa was dry (Roberts and Taieb, 1993).
```

or

```
The Younger Dryas in Africa was dry (Roberts & Taieb, 1993).
```

depending on the format that each journal defines in its guidelines. In many cases more than one reference is cited after a scientific statement, such as

```
The Younger Dryas in Africa was dry (Williamson et al., 1993;
Roberts et al., 1993).
```

with the references listed chronologically, with the oldest first. Bearing in mind the space limitations of most journals, however, we need to decide which references to actually cite for a statement, which brings us back to the original topic of this section. As an example, the maximum number of references in the main text in Nature and Science is thirty or fifty citations, depending on the type of paper, whereas there is no limit to the number of references in the electronic supplement.

2.5 Extracting Text, Data and Graphs from Literature

Having identified the most relevant article on a specific topic, we now wish to extract scientific information from that article, not only by reading and remembering that information, but also by taking text and figures from the electronic document for incorporation into a poster, a presentation slide or any other type of presentation, of course within the limits of copyright regulations. Most posters and presentations contain original research and the figures are therefore newly-designed by the researcher. In the bachelor's level course that inspired this book, however, the students were not conducting their own research and therefore prepared posters, abstracts, and presentations based on previously published work by other researchers. University lecturers, however, cover broader topics, usually complemented by their own original research.

The level of modification that is possible for text and figures extracted from a published manuscript or book, prior to its incorporation into a presentation, depends on the way the document has been electronically saved. Computer graphics are generally stored and processed as either vector or raster data. Most of the data types encountered in the following chapter on processing geoscientific information are vector data, i.e., points, lines and polygons. Drainage networks, the outlines of geologic units, sampling locations and topographic contours are all examples of vector data. In Chapter 6, coastlines are stored in a vector format while bathymetric and topographic data are saved in a raster format. Vector and raster data are often combined in a single data set, for example in order to display the course of a river on a satellite image. Raster data are often converted to vector data by digitizing points, lines or polygons. Conversely, vector data are sometimes trans-

formed to raster data, which requires a degree of discretization and gener-
alization. Images are generally presented as raster data, i.e., as a 2D array of
color intensities. Images are everywhere in geosciences. Field geologists use
aerial photographs and satellite images to identify lithologic units, tectonic
structures, landslides, and other features within a study area. Petrologists
and micropaleontologists use microscope images of minerals and fossils
from either an optical microscope or a scanning electron microscope.

The PostScript (PS) format is used to store scalable vector graphics, raster images, and text. It was developed by John Warnock at PARC, the Xerox
research institute. Warnock was also co-founder of Adobe Systems, where
the EPS format was created. The PostScript vector format would have never become an industry standard without Apple Computers. In 1985 Apple
needed a typesetter-quality controller for the new Apple LaserWriter printer
and the Macintosh operating system, and adopted the PostScript format.
The third partner in the history of PostScript was the Aldus company, the
developer of the PageMaker software and now a part of Adobe Systems.
The combination of Aldus PageMaker software, the PS format, and the
Apple LaserWriter printer led to the creation of Desktop Publishing. The
Encapsulated PostScript (EPS) format was then developed by Adobe Systems
as a standard file format for importing and exporting PS files. Whereas a PS
file is generally a single-page format containing either an illustration or a
text, the purpose of an EPS file is to allow the inclusion of other pages, i.e., it
is a file that can contain any combination of text, graphics and images. The
Portable Document Format (PDF) designed by Adobe Systems in 1993 is
now a true, self-contained, cross-platform document.

Most journal articles and electronic books are today available as PDF
files, which may include editable text in a vector format, vector graphics,
and raster images. The PDF file of the article by Neil Roberts, however, is a
scanned document in raster format, as is revealed by zooming in on the text
and figures. It is therefore not possible to extract text and figures in a vector
format for further modification. Text recognition, automated vectorization,
or manual digitizing can partly solve this problem, and pixel graphics could
also be extracted without further editing and overlaid by vector additions in
a separate layer (see Chapter 9).

More recent articles published in Nature, Science, or any other journal
are available in a vector format. As an example, we use my own article

Trauth MH, Maslin M, Deino MA, Deino A, Strecker MR (2005) Late
Cenozoic Moisture History of East Africa. Science 203:2051-2053.

published in 2005 in Science (http://www.sciencemag.org). As with Nature, the Science webpage has a search field in which we can enter the title *Late Cenozoic Moisture History of East Africa* and quickly find the article.

```
Late Cenozoic Moisture History of East Africa
Martin H. Trauth, Mark A. Maslin, Alan Deino, and Manfred R.
Strecker
Science 23 September 2005: 2051-2053.Published online 18 August
2005 [DOI:10.1126/science.1112964]
Abstract Full Text Full Text (PDF) Supporting Online Material
```

As well as the article being available as the abstract only, as the full text online, or as the full text in a PDF, it also includes supporting online material. Once again, the abstract is freely available:

```
Lake sediments in 10 Ethiopian, Kenyan, and Tanzanian rift basins
suggest that there were three humid periods at 2.7 to 2.5 million
years ago (Ma), 1.9 to 1.7 Ma, and 1.1 to 0.9 Ma, superimposed on
the longer-term aridification of East Africa. These humid periods
correlate with increased aridity in northwest and northeast
Africa and with substantial global climate transitions. These
episodes could have had important impacts on the speciation
and dispersal of mammals and hominins, because a number of key
events, such as the origin of the genus Homo and the evolution of
the species Homo erectus, took place in this region during that
time.
```

It provides the most important information and conclusions from the article. Access to the full text, whether online or as a PDF file, requires a subscription to the journal, or alternatively it can be accessed via a pay-per-view option. Having downloaded the PDF file, we quickly notice that it is in a true vector format since, for instance, we are able to mark and copy text fragments and to paste them into word processing software. Differences exist, however, in the way a PDF reader deals with the text, in particular with respect to indented text, special characters, and paragraph styles. Using Adobe Acrobat on a computer running the Mac OS X operating system, for instance, right justification in the linefeeds of text is interpreted as a paragraph end marker at the end of each justified line, both in the abstracts and in the main text. For instance, the first paragraph of the main text appears as

```
Recent investigations of both terrestrial and
marine paleoclimate archives have led to a
concerted debate regarding the nature of Late
Cenozoic environmental changes in East Africa
and their influence on mammalian and hominin
evolution (1-3).
```

In this case, the paragraph end markers have to be removed either manually or using search and replace, prior to making further modifications of the text. Making the same experiment using Apple Preview avoids this problem, resulting in

```
Recent investigations of both terrestrial and marine paleoclimate
archives have led to a concerted debate regarding the nature of
Late Cenozoic environmental changes in East Africa and their
influence on mammalian and hominin evolution (1-3).
```

Another problem when extracting text fragments from an article is the incorrect transfer of subscripts, superscripts, and other paragraph styles. For instance, the term $^{40}Ar/^{39}Ar$ *age calibration* from the second paragraph of the main text becomes

```
(...)
with intercalated volcaniclastic deposits that
permit high-precision 40Ar/39Ar age calibration
of lake-level highstands (5, 6) (Fig. 2).
(...)
```

in which the numbers *40* and *39* are transferred as a smaller font but not as superscripts, with both Adobe Acrobat and Apple Preview. Copying this paragraph from the PDF viewed with Apple Preview again interprets the linefeeds correctly but puts hyphens and a space within words at linefeeds, such as *calibra- tion* in the text below:

```
Although much smaller than the lakes in the western branch and
often subaerially exposed, these basins host a rich sedimentary
record, with intercalated volcaniclastic deposits that permit
high-precision 40Ar/39Ar age calibra- tion of lake-level
highstands (5, 6) (Fig. 2).
```

These errors again need to be corrected either manually or by using the search and replace option in text processing software.

In the next experiment we try to import the two figures included in the article by Trauth et al. (2005). We use the import feature of vector graphics software such as the open source *Inkscape* software, or the commercial *Adobe Illustrator* software. Opening the first page of the PDF file reveals that Figure 1, which shows a map of East Africa, is indeed in a vector format. After importing the figure we can click on individual elements such as polygons or text, and modify colors, line thicknesses and font sizes (see Chapter 8). However, importing the second page of the article, which includes Figure 2, reveals that this figure is in a pixel format, preventing any further modification. We suspect that the publisher may include data graphs, such as the compilation of lake records presented in Figure 2, in a pixel format in order to prohibit unwanted modification or manipulation of the author's original graphs. This figure, however, could be imported into pixel software such as *Gimp* or *Adobe Photoshop* and modified with the tools available for editing photos and other pixel graphics. Furthermore, the pixel graphics can be imported into vector graphics software and overlaid with vector elements such as polygons and text.

2.6   Organizing Literature in a Computer

Familiarization with a new topic at the beginning of a research project requires the collection and reading of a large number of articles, books, and reports, and their organization as PDF files on a computer. To organize the collection of references and electronic documents, we need literature or bibliographic management software. A large number of both open source and commercial software products are available for managing your literature. We use the open source *BibDesk* (http://bibdesk.sourceforge.net) software for Mac OS X as well as the commercial *EndNote* (http://www.endnote.com) and *Papers* (http://mekentosj.com/papers), which are representative examples of the typical design of literature management software products.

Bibliographic management software typically consists of a searchable reference data base, often linked with online data bases such as Google Scholar and Thomson Reuters ISI Web of Science, and a file directory service for managing the PDF files on a hard drive. Most software packages can be integrated into word processing software such as *Microsoft Word* or *Apple Pages*, and reference lists are then generated automatically in a specific journal's layout from references included in the manuscript text. Alternatively, software based on the *BibTeX* standard, such as *BibDesk* for the Unix-based Mac OS X, or *KBibTeX* for Linux or Unix, uses *LaTeX* to prepare formatted literature lists.

The most popular commercial software for organizing reference data bases is EndNote. The software was first released in 1988 and is therefore one of the oldest reference management tools, designed before Adobe had released the PDF. The software was therefore primarily created to manage citation libraries rather than to organize PDFs on a hard disk. In the early days, the user had to type references into the data base manually, using an input mask with fields for authors, title, year and other kinds of information related to the article. Subsequently, most literature search data bases allowed users to export the search results in a format that could be directly imported into EndNote. Today, EndNote can also be used to organize PDF files. The software can be integrated with word processors and automatically generates reference lists in the format required by a specific journal.

Most universities provide campus licenses that can be acquired at a reasonable price. EndNote is available for Microsoft Windows and Mac OS X operation systems.

Launching the EndNote software brings up a single multi-panel user interface that includes a panel for browsing the library (left panel), a list of all references in the data base (upper right panel) and a panel with tabs for *Preview*, *Search*, and *Quick Edit* (lower right panel). Since EndNote is linked to Thomson Reuters ISI Web of Science, we can again start a search for the keywords *Younger Dryas East Africa*. We select *Web of Science* from *Online Search* in the *My Library* panel, select Title as the criterion for the search in the *Online Search* tab, type the keywords *younger dryas east africa* in the search field, and then press *Search*. This yields a list of several references including the article by N. Roberts, which is the earliest article listed. Double-clicking the article in the list opens a new window displaying the data sheet for this particular reference with the categories *Author, Year, Journal, Volume, Issue, Pages*, and so forth. The data sheet also includes the abstract, number of citations, and the author's address. The *Reference* menu of the software includes a *Find full text* feature that automatically includes the PDF of the article if the user has an active subscription to Nature. Having included the reference and the file of the article in the data base, we can export the reference in a format of the user's choice for inclusion in a manuscript requiring a specific format. As an example, the reference to the article by N. Roberts can be previewed and exported in the Nature format:

1 Roberts, N. *et al.* Timing of the Younger Dryas Event in East-Africa from Lake-Level Changes. *Nature* **366**, 146–148 (1993).

Switching to the format of Elsevier's Earth and Planetary Science Letters journal creates the reference

Roberts, N., Taieb, M., Barker, P., Damnati, B., Icole, M., Williamson, D., 1993. Timing of the Younger Dryas Event in East-Africa from Lake-Level Changes. *Nature* **366**, 146–148.

The format for the Proceedings of the National Academy of Sciences of the United States of America (PNAS) is as follows:

1. Roberts N, *et al.* (1993) Timing of the Younger Dryas Event in East-Africa from Lake-Level Changes. *Nature* **366**(6451):146–148.

The user interface and features of the free BibDesk software are very similar to those of EndNote. The open source software is a BibTeX front end that uses LaTeX to export formatted reference lists. BibDesk (http://bibdesk.sourceforge.net) is only available for Mac OS X; it was first released in 2002 and can be downloaded as a separate product, without LaTeX. A recommended alternative is to download the software as part of the *MacTeX* bundle (http://www.tug.org/mactex), which includes the official standard distribution of *TeX* for Macs. Installing MacTeX allows formatted reference lists to be prepared without requiring a separate LaTeX editor. We can select *Web of Science* from the *Searches* menu, causing a search panel to open in which entering the keywords *Younger Dryas East Africa* yields no results. Since there are no further options available for the search fields, we need to consult the ISI Web of Science help page for further instructions. We learn that we have to include the search terms manually and modify our search into *TS=Younger Dryas East Africa*, where *TS* stands for *Topic Search*. This again yields the papers by D. Williamson and N. Roberts as the earliest references on the list. The data sheet for the article by N. Roberts includes a link to the ISI Web of Science webpage but no direct link to the PDF of the article on the webpage of the journal. The *Preferences* menu provides numerous BibTeX styles with which to format the reference list for export. Choosing *apalike* as an example format and then clicking the *TeX Preview* button from the user interface results in the formatted output

ROBERTS, N., TAIEB, M., BARKER, P., DAMNATI, B., ICOLE, M., and WILLIAMSON, D. (1993). Timing of the younger dryas event in east-africa from lake-level changes. *Nature*, 366(6451):146–148.

which follows the capitalized format used in the ISI Web of Science data base for the authors' names. The commercial EndNote software clearly corrects this mistake and results in the output seen previously.

An award-winning commercial alternative to EndNote and BibDesk for Mac OS X is Papers by *Mekentosj* from The Netherlands (Fig. 2.5). The software was released in 2007 and is currently available as Version 2, which provides numerous additional features to those of the first version. In contrast to EndNote and its open source counterpart BibDesk, this software was designed from the start to maintain large libraries of PDF documents comprising journal articles, books, theses, and other types of scientific literature. As with the other tools, Papers offers repository searches using the ISI Web of Science and other data bases, editing of the file metadata using data sheets, full-screen reading, and commenting, as well as multiple import and export features. A great advantage of the software is the very straight-forward approach to renaming and reorganizing the PDF files on a hard drive. As soon as a PDF file has been imported into the software, it is renamed in a way defined by the user. In our example, for instance, the PDF file of the article

Fig. 2.5 Screenshot from *Papers 2.1* by *Mekentosj* (http://www.mekentosj.com), an award-winning bibliographic management software for organizing a collection of references and electronic documents.

by N. Roberts is renamed *Nature 1993 Roberts.pdf* and copied into the *1993* folder and a *Roberts* subfolder in the *Papers* directory, in the user's home directory. This contrasts with the way EndNote organizes its Library as it first creates a directory named *EndNote Library.Data*, then a subdirectory

SEARCHING AND REVIEWING SCIENTIFIC LITERATURE

2

named *Library Trauth.Data*, in which the file named *Library Trauth.enl* is located. The directory *EndNote Library.Data* also contains a subdirectory, among others, named *PDF*, which in turn contains the folder *Roberts-1993-Timing of the Younge-2052588800*. This folder, with its long name, contains the PDF renamed as *Roberts-1993-Timing of the Younge.pdf*.

BibDesk, EndNote, and Papers are only three of the reference management tools that are available for different operating systems, but they are representative examples. All of these tools help in managing rapidly expanding reference data bases and PDF collections during the course of a typical earth sciences research project.

Recommended Reading

Bradley RS (1999) Paleoclimatology – Second Edition: Reconstructing Climates of the Quaternary. Academic Press, Burlington

Hirsch JE (2005) An index to quantify an individual's scientific research output. Proceedings of the National Academy of Sciences 102:16569-16572

Langenberg H (2010) Editorial – Number crunch. Nature Geoscience 3:445.

Roberts N, Taieb M, Barker P, Damnati B, Icole M, Williamson D (1993) Timing of the Younger Dryas event in East Africa from lake-level changes. Nature 366:146–148

Ruddiman WF (2000) Earth's Climate – Past and Future. W.H. Freeman and Company, New York

Williamson D, Taieb M, Damnati B, Icole M, Thouveny M (1993) Equatorial extension of the Younger Dryas Event - Rock Magnetic Evidence From Lake Magadi (Kenya). Global and Planetary Change 7:235–242

3 Internet Resources for Earth Science Data

3.1 Introduction

This chapter deals with electronic data resources, searching for specific data, transferring data between servers and computers (*clients*), and data storage. Section 3.2 introduces the systems available for data storage, starting with a historical retrospective and ending with the most recent storage systems such as DVDs, CDs, and flash memory. Section 3.3 discusses the various data and file formats used. These include ASCII and binary file formats, generic binary file formats such as the JPEG format for images or the PS formats for vector graphics, and software-specific file formats (such as the Microsoft Word file DOC format or the MATLAB binary MAT format). Section 3.4 explains how data are transferred from one computer to another via data networks: it provides overviews of the history of data transfer and of the established transfer techniques and protocols. The remaining sections in this chapter go on to present several examples of data resources and searches, which are then used in subsequent chapters for the visualization and presentation of geoscientific information (Sections 3.5 to 3.8). Section 3.9 concludes the chapter with methods for storing and organizing data in a computer.

3.2 Data Storage in a Computer

Humans first started using technology to store information outside their brains about 2.5 million years ago. At that time the first *stone tools* were made, and one human individual was therefore able to pass on the technique for making things without the use of words. There have been many developments since that time, including the evolution of language and, some time later, of writing. The love song of the princess of the Sumerian king Shu-Shin, dating from about 5,500 years before the present, is the earliest known writing in history. Early hieroglyphs were written on *clay* or *limestone tablets*, and later on *papyrus*, from which the Greek word $\beta\iota\beta\lambda o\varsigma$ or

M. H. Trauth and E. Sillmann, *MATLAB® and Design Recipes for Earth Sciences*,
DOI: 10.1007/978-3-642-32544-1_3, © Springer-Verlag Berlin Heidelberg 2013

biblos (meaning book), and the English word *Bible*, are derived. Bibles were for many years copied by ancient scribes, and these copies were then copied over and over again, including notes on the margins of the pages as well as many errors, until *typography* was invented in China in the eleventh century, and subsequently in Europe (by Johannes Gutenberg) in the mid-15th century. Copying books manually was then no longer necessary, nor were public readings of books in lecture halls.

Portable information has become increasingly compressed over the last two centuries through, for instance, the invention of microfilm in the mid-19th century, and the punched cards originally used for textile looms and fairground organs. Punched cards were used with computers until the mid 1980s, when they were replaced by magnetic tapes. Data on tapes are stored using a binary system, with the symbol *0* for *no hole* (using the terminology from punched cards) and *1* for *hole*. Tapes were then largely replaced by diskettes or *floppy disks*, invented by IBM and available in 8 inch, 5¼ inch and 3½ inch sizes making the winding of tapes no longer necessary. Of course the life span of these magnetic media (about 5–30 years) is much shorter than that of clay tablets (up to 12,000 years). On the other hand, there is no comparison between the 3.2 MB maximum storage capacity of a 3½ inch floppy disk (corresponding to about 11 million zeros or ones) and the very limited amount of space available on clay tablets.

The first *hard drive disks* were introduced by IBM in 1956, but the relationship between cost and storage capacity prevented them from being attractive for the mass market until the 1980s. The storage capacities of hard drive disks today can be up to 3 TB. Medium-capacity *magneto-optical* (MO) *drives*, and the more affordable *ZIP drives* (100 MB, later 250 MB and even 750 MB), became very popular in the 1990s but were later replaced by the much cheaper rewritable *compact disks* (CDs). The CD was introduced by Sony and Philips Consumer Electronics in the late 1970s as an optical storage medium for music, and was restricted to a maximum storage capacity of 870 MB. The *Digital Versatile Disk* (DVD) was introduced by a consortium of Sony, Philips, Toshiba, and Time Warner in 1995 as a medium with greater storage capacity (from 4.7 to about 17 GB) than a CD, but with similar dimensions. The lifespan of data stored on CDs and DVDs is at least 10 years, although manufacturers claim at least 30 years.

A *flash memory* on computer storage chips, which can be electronically erased, was introduced by Toshiba in the 1980s and soon found its way into consumer products such as digital cameras, digital audio players, mobile phones, and more recently, even into laptops, as a very robust and fast storage medium. The maximum storage capacity of flash memory is today 32 GB but several chips can be bundled together in laptops and mobile phones to increase the total amount of memory available.

3.3 Data Formats in Earth Sciences

A computer generally stores data as *binary digits* or *bits*. A bit is analogous to a two-way switch with two states, on = 1 and off = 0. The bits are joined together to form larger groups, such as bytes consisting of 8 bits, in order to store more complex types of data. Such groups of bits are then used to encode data such as numbers or characters. Unfortunately, different computer systems and software use different schemes for encoding data. For instance, the characters in the widely-used text processing software Microsoft Word differ from those in Apple Pages. Exchanging binary data is therefore difficult between different computer platforms and software. Binary data can be stored in relatively small files if the researchers processing the files are using similar systems for data exchange. The transfer rate for binary data is generally faster than that for other file formats.

A number of different formats for exchanging data have been developed during recent decades. The classic example of a data format that can be used with different computer platforms and software is the *American Standard Code for Information Interchange* (ASCII), which was first published in 1963 by the American Standards Association (ASA). As a 7-bit code, ASCII consists of $2^7 = 128$ characters (codes 0 to 127). Whereas ASCII-1963 lacked lower-case letters, they were included in the ASCII-1967 update, together with various control characters, such as *escape* and *line feed*, and various symbols, such as brackets and mathematical operators. A number of variants have since appeared in order to facilitate the exchange of text written in non-English languages, such as the expanded ASCII, which contains 255 codes, e.g., the Latin-1 encoding.

In 1987, Microsoft Corporation introduced the cross-platform 8-bit *Rich Text Format* (RTF), which includes *escape sequences* allowing the text to be formatted using a *WYSIWYG* (*what you see is what you get*) type of text processor (http://microsoft.com). Binary DOC files generated by *Microsoft Word* (recently replaced by DOCX files with XML support), ODT files generated by *OpenOffice Text*, and *pages* files generated by *Apple Pages*, contain much more formatting information than RTF files but at the expense of being less compatible. *Apple Pages*, however, contains converters for reading DOC files and converting them into *pages* files, but the latest 2011 release of *Microsoft Word* cannot read *pages* files. The same applies to exchanges between *Apple Numbers* files and *XLS* files generated by

Microsoft Excel, and between Apple Keynote files and *PPT* files generated by *Microsoft PowerPoint*.

Similar incompatibilities exist between graphics software packages, such as between *Adobe Illustrator* and *Photoshop*, or *CorelDraw* and *Canvas* (as commercial products), and between *Inkscape* and *Gimp* (as open source products). Computer graphics are stored and processed as either *vector data* or *raster data*. Many different kinds of plots, such as line graphs or bar plots, and also more complex graphics such as those displaying drainage networks, the outlines of geologic units, sampling locations, or topographic contours, are all examples of vector data. Vector data use mathematical equations to describe geometrical primitives such as polygons and points. The use of these mathematical equations, basically describing the relationship between xy coordinates of polygons, makes it possible to modify the polygons in many ways, e.g., by editing the vertices, scaling the entire polygon, or by changing line thicknesses, colors or other properties of the polygon (see Chapters 5 and 6). Images, however, are generally represented as raster data, i.e., as a 2D array of color intensities. The *raster data* are stored as 2D arrays of elements, each of which contains information on the color intensity values of the *pixels* (short for *picture elements*) (see Chapter 7).

Numerous formats are available for saving vector and raster data into a file, each of which has its own particular advantages and disadvantages. The choice of one format over another in an application depends on the way the images are to be used in a project, and whether or not the images are to be analyzed quantitatively. The *Compuserve Graphics Interchange Format* (GIF) was developed in 1987 for raster images using a fixed colormap of 256 colors. The GIF uses compression without any loss of data and was designed for fast transfer rates over the Internet. The limited number of colors means that it is not the right format for the smooth color transitions that occur in aerial photos or satellite images. It is, however, often used for line art, maps, cartoons and logos (http://compuserve.com).

The Microsoft Windows *Bitmap Format* (BMP) is the default image format for computers using the Microsoft Windows operating system. However, numerous converters also exist for reading and writing BMP files on other platforms. Various modifications of the BMP format are also available, some of them without compression and others with effective and rapid compression (http://microsoft.com). The *Tagged Image File Format* (TIFF) was designed by the Aldus Corporation and Microsoft in 1986 and went on to become an industry standard for image-file exchange. A TIFF file includes an image file header, a directory, and the data, in all available graphics and image file formats. Some TIFF files even contain vector and raster versions of the same picture, as well as images at different resolutions and with different colormaps. The main advantage of TIFF files was originally their portability. A TIFF should perform on all computer platforms but, unfortunately, the numerous modifications of the TIFF that evolved in subsequent years often resulted in incompatibilities, to the extent that the TIFF is now often referred to as *Thousands of Incompatible File Formats*.

The *Joint Photographic Experts Group* (JPEG) was established in 1986 for the purpose of developing various standards for image compression. Although JPEG is an acronym for the committee name, it is now widely used as the name for an image compression and a file format. This compression consists of grouping pixel values into 8×8 blocks and transforming each block using a discrete cosine transform. As a result, all unnecessary high-frequency information is deleted, which makes this form of compression irreversible. The advantage of the JPEG format is the availability of a three-channel, 24-bit, true color version. This allows images with smooth color transitions to be stored. The new *JPEG-2000* format uses a wavelet transform instead of the cosine transform (Section 5.8) (http://jpeg.org).

The *PICT* format was developed by Apple Computers in 1984 as the default format for Macintosh graphics. It can be used for both raster images and vector graphics. PICT uses various methods for compressing data. The PICT 1 format only supports monochrome graphics, but PICT 2 supports a color depth of up to 32 bits. The PICT format is not supported on other platforms although some PC software tools can work with PICT files (http://apple.com). The *PostScript* (PS) format and the *Encapsulated PostScript* (EPS) format have already been discussed in Chapter 2. An EPS file is designed to allow the inclusion of different PS files, thus allowing text, graphics, and images to be combined. The *Portable Document Format* (PDF) designed by Adobe Systems is now a truly self-contained cross-platform document. PDF files, also previously discussed in Chapter 2, contain the complete formatting of vector illustrations, raster images and text, or a combination of all these, including all necessary fonts (http://adobe.com).

Most computer operating systems and software today allow the import and export of vector and raster graphics in most of these formats, although minor incompatibilities still exist, which can often be time consuming. Binary AI files generated by Adobe Illustrator or PSD files by Adobe Photoshop contain a far greater amount of editing information than other formats, but at the expense of being less compatible.

In addition to these text, table, and graphics formats, various data compression and archive formats permit multiple files to be bundled together into a single archive file, and the file size to then be compressed. The ZIP format was introduced by Phil Katz in 1989 and is now used to compress data and software files for faster file transfer over the Internet. The TAR format is the corresponding method for file compression in the Unix world, first introduced in 1988. Both ZIP and TAR are often used together to create GZIP files. Most computer operating systems contain software to compress and uncompress ZIP and TAR files, as well as or other compressed files.

The next section discusses the transfer of data through the Internet, while subsequent sections illustrate the exchange and processing of geoscientific information in various data formats.

3.4　Data Transfer between Computers

The Internet has dramatically improved scientific communication since the early 1990s. Its history extends back into the 1960s and early 1970s, when it was first established as the military communication network *Advanced Research Projects Agency Network* (ARPANET). Between the mid 1970s and the early 1980s, university networks were also connected, commissioned by the United States National Science Foundation (NSF), and then opened to other networks in 1988. The connection of individual networks was possible through the use of the standard Internet protocol suite *Transmission Control Protocol/Internet Protocol* (TCP/IP). The TCP/IP protocol defines the format of digital information transferred through a network and the rules for such transfers between computer hardware and networks, including authentication and error management. The use of the TCP/IP protocol since the 1970s, the commercialization of the network in the late 1980s, and the establishment of the World Wide Web (WWW) in the early 1990s, have facilitated in the enormous growth of the Internet, with over two billion users worldwide in 2011. A comprehensive historical summary of the Internet can be found at http://en.wikipedia.org/wiki/Internet.

While the TCP/IP protocol is the main protocol for the Internet as a whole, it operates in concert with several other protocols that are used for communication between computers, for electronic mail services, for the transfer of large data sets, or for accessing the World Wide Web. One of the first network protocols was *Telnet*, which provided bidirectional communication between computers and servers using either a *text terminal* or console connection. The applications running a text terminal are command-line interpreters or shells, such as the *Command Windows* in MATLAB, allowing commands or series of commands to be entered after a prompt, which are then executed after pressing *Enter* (see Chapter 4.2). Examples of text terminal applications are *Terminal* and *Console* in Unix-based operating systems such as Solaris, Linux, or Mac OS X, and *cmd.exe* or *PowerShell* in Microsoft Windows.

After launching the terminal software, a typical Telnet session starts with typing the telnet command followed by the *Internet Protocol* (IP) address or the hostname of the remote computer. The numerical IP address and human-memorable hostname are linked by the entries in a lookup table stored on a name server, typically of the *Domain Name System* (DNS) type. As an example, we can connect to the communication server of the University of Potsdam by typing the IP address

```
telnet 141.89.68.1
```

after the terminal prompt > or the hostname

```
telnet persius.rz.uni-potsdam.de
```

The University of Potsdam communication server responds with

```
Trying 141.89.68.1...
Connected to persius.rz.uni-potsdam.de.
Escape character is '^]'.
SunOS 5.9

              Willkommen am Kommunikationsserver
                   der Universitaet Potsdam!

              persius.rz.uni-potsdam.de [141.89.68.1]

         Bei Problemen: Tel: +49-331-977-1792(1328)
login:
```

requiring a login name and password. If you do not have a user account with the University of Potsdam's communication server, you can connect to your own university communication server instead. After logging in you will see a welcome message and the communication server prompt. You can then interact with the server using Unix commands, some of which are used later in Chapter 4.2 while working with the MATLAB shell or Command Window. A list of Unix commands can be found at

```
http://en.wikipedia.org/wiki/Unix_commands
```

As an example,

```
ls
```

simply lists the directory contents, i.e., the files stored in the user's home directory. The command

```
cd foldername
ls
```

allows you to move into a specific directory and again lists the contents. The Telnet protocol is, however, not encrypted, and hacking activities on the Internet have now made it necessary to use encrypted alternatives. In fact, during the writing of this book, the computer center of the University of Potsdam has disabled Telnet, so that we now get the response:

```
Trying 141.89.68.1...
telnet: connect to address 141.89.68.1: Connection refused
telnet: Unable to connect to remote host
```

The most popular alternative is the *Secure Shell* (SSH) network protocol, introduced in 1995, which sends information, in particular passwords, in an encrypted format. The command to log in to a remote computer system using a terminal's software is

```
ssh username@persius.rz.uni-potsdam.de
```

or

```
ssh username@141.89.68.1
```

causing the server to respond with the welcome message and a prompt.

In the early 1990s, before the introduction of specialized software for handling electronic mail, emails were handled using command-line tools or the text user-interface-based *Pine* email software introduced by the University of Washington in 1989. The *Simple Mail Transfer Protocol* (SMTP) for email transport across the Internet was introduced in the early 1980s by the University of California Berkeley, and its successor *Sendmail X* is still the standard protocol for emailing. Today, email client software such as Microsoft *Outlook* and Apple *Mail*, which are built into their respective operating systems, or the free and very popular alternatives Mozilla *Thunderbird* (http://mozilla.org/thunderbird) or Qualcomm *Eudora* (http://eudora.com), facilitate the management of email. The exchange of emails between servers and personal computers is managed either by the *Post Office Protocol* (POP, now in the POP3 version) or the *Internet Message Access Protocol* (IMAP).

The *File Transfer Protocol* (FTP) is used to transfer larger packages of data from one computer to another over a TCP-based network. The protocol was written in the early 1970s by Abhay Bhushan, who was also involved in designing the TCP/IP Internet architecture. We can use the FTP service on the University of Potsdam's communication server, or any other communication server, to demonstrate the use of file transfer services by first typing

```
ftp username@persius.rz.uni-potsdam.de
```

after the prompt requiring you to log in with a password. After the welcome message, we can type

```
ls
```

to get the directory contents and

```
get filename
```

to download a specific file to our local computer system. Most servers also provide an SSH or *Secure File Transfer Protocol* (SFTP) using the SSH protocol for data encryption, which can be obtained by typing

```
sftp username@persius.rz.uni-potsdam.de
```

Today, FTP and SFTP are often provided by Internet browsers when accessing larger files for download, with their use being obvious in the change from *http://* to *ftp://* in the software's address bar. The *http* in the address bar stands for the *Hypertext Transfer Protocol* (HTTP), which was introduced in 1991 for communicating on the *World Wide Web* (WWW). Hypertext is text with references (i.e., *hyperlinks*, or more simply, *links*) that allow the user to immediately access information related to the text, images, or any other type of information shown on a webpage. Webpages, including the server and client software (*browsers*), are written in the *HyperText Markup Language* (HTML) developed at the European Organization for Nuclear Research (CERN) in 1990 to provide and access webpages on the Internet. These webpages are identified by their *Uniform Resource Locator* (URL) in the address field of the browser. For example, the URL for the relevant article on Wikipedia is

```
http://en.wikipedia.org/wiki/Uniform_Resource_Locator
```

The World Wide Web today combines most of the Internet services described within this section, including mail, browsing remote directories, and file transfer. It also provides the ability to access and view multimedia information such as ebooks, music, and videos, as well as cloud and similar network services, thereby making a significant contribution to the accessibility of scientific information.

3.5   Internet Resources: When was the Younger Dryas?

In Chapter 2 we searched for the most relevant journal articles on the influence of the Younger Dryas in East Africa and found this paper by N. Roberts:

```
Roberts N, Taieb M, Barker P, Damnati B, Icole M, Williamson
D (1993) Timing of the Younger Dryas event in East Africa from
lake-level changes. Nature 366:146-148
```

We used the popular web-based, open source encyclopedia *Wikipedia* to learn that the Younger Dryas stadial was a geologically brief (1,300 ± 70 years) cold-climate period between approximately 12,800 and 11,500 years ago (between 10,800 and 9500 BC) (see http://en.wikipedia.org/wiki/Younger_Dryas). The two papers cited as sources in the Wikipedia article were by W.H. Berger (1990) and R. Muscheler (2008).

Let us assume that we are working on a review article on the worldwide influence of the Younger Dryas and we wish to plot a time series from several different locations on Earth, covering the critical time interval. Ice cores retrieved from Greenland in the early 1990s have made a considerable contribution to our understanding of the exact onset, duration, and termination of the Younger Dryas cold interval. A review article on these ice core results, published in the year 2000, was included in the list of titles from the Google Scholar search (Section 2.3).

```
4. RB Alley (2000) The Younger Dryas cold interval as viewed
from central Greenland, Quaternary Science Reviews 19:213-226
[Citations: 218]
```

Let us obtain the ice core data described in the article by R.B. Alley. We enter *Younger Dryas Ice Core* in the search field of

```
http://google.com
```

and obtain (in early 2011) a list of about 34,100 results. One of the first results on the list is

```
NOAA Paleoclimatology Program - Alley 2000 Greenland Ice Core
Data
10 May 2004 ... The Younger Dryas cold interval as viewed from
central Greenland ... Greenland ice-core records provide an
exceptionally clear picture of ...
www.ncdc.noaa.gov/paleo/.../alley2000.html
```

which appears to contain the data described by R.B. Alley (2000) on a webpage of the *National Climatic Data Center* (NCDC), which is part of the *National Oceanic and Atmospheric Administration* (NOAA), on

```
http://ncdc.noaa.gov/paleo/pubs/alley2000/alley2000.html
```

The webpage provides a the title and author of the article, and the most important graph (showing the temperature and snow accumulation time series over the last 20,000 years from the GISP2 ice core), as well as the abstract, the cited references, and a link to the Elsevier *Science Direct* website. We click on *Data* to obtain the ice core data shown on the graph, on

```
ftp://ftp.ncdc.noaa.gov/pub/data/paleo/icecore/greenland/summit/
gisp2/isotopes/gisp2_temp_accum_alley2000.txt
```

Depending which internet browser we are using, we may be required to log on as a guest before being directed to the text file. We notice that the protocol changes from HTTP to FTP, as indicated by the *Uniform Resource Locator* (URL) in the address field. The data webpage seems to be in an ASCII text format, also indicated by the name of the corresponding file containing the data *gisp2_temp_accum_alley2000.txt*. We save this file on the hard drive of our computer and open it in a text editor such as Apple *TextEdit*, Microsoft *WordPad*, *TextWrangler* (free sofware from http://barebones. com/products/textwrangler that allows line numbers to be shown), or the MATLAB *Editor* (which also provides line numbering). The first 70 lines of the file provide extensive information on the data set, the source, a description of the data, and the article to be cited if the data are used in another article or in a textbook. The file header also indicates that the file was updated a few years after publication of the corresponding article. Lines 75 to 1707 contain the temperature data in the second column (in degrees C) and the corresponding ages in the first column (in thousands of years before present), while lines 1717 to 3414 contain the snow accumulation data in the second column (in meters per year) and as before, the corresponding ages in the first column, from the Greenland ice core. Comparing the age columns for both data sets reveals differences in the ages for the temperature and the snow accumulation data points. We cannot therefore copy both data sets into a single table with a single column for the ages. Instead we mark, copy, and paste the two data sets separately into two files that we call *icecore_temperature_data.txt* and *icecore_snowaccumulation_data. txt*. The temperature data file *icecore_temperature_data.txt* includes one header line and 1632 lines of data:

```
        Age              Temperature (C)
     0.0951409             -31.5913
     0.10713               -31.622
     0.113149              -31.6026
     (cont'd)
```

The snow accumulation data file *icecore_snowaccumulation_data.txt* includes one header line and 1697 lines of data:

```
      Age            Accumulation
  0.144043             0.244106
  0.172852             0.246155
  0.20166              0.248822
  (cont'd)
```

In Chapter 5 we will use these data sets as examples to create plots with MATLAB. Alternatively, we could also use spreadsheet applications such as Microsoft *Excel*, Apple *Numbers* or the OpenOffice *Spreadsheet* to import, manipulate, and visualize the data. As an example, we will now demonstrate how to process the data in *icecore_temperature_data.txt* using the OpenOffice Spreadsheet.

After launching the OpenOffice software, we select *Spreadsheet* to create a spreadsheet document. We then import the GISP2 temperature data using the *Insert* menu by selecting *Sheet from File*, and then selecting the file *icecore_temperature_data.txt* from the working directory on the hard drive. The *Insert Sheet* and *Text Import* dialogs appear and show a preview of the imported table. Since the original text file uses different numbers of spaces as separators in each individual row, we have to select the option *Merge delimiters* in the *Text Import* window, and then press *OK* twice to get the expected results as a two-column spreadsheet, including the headers Age and Temperature. Selecting the *Format* menu, and then *Cells ...*, allows us to define the format of the two columns, e.g., by choosing the category *Number*, the floating point representation *–1234.12*, and 4 decimal places. We then save the spreadsheet as *icecore_temperature_data.ods* on the hard drive of our computer.

The data can be graphically displayed by selecting both columns and then selecting *Chart* from the *Insert* menu and choosing the type of chart that we want to use. We decide to display the data as an *XY (Scatter)* plot with the option *Lines only*, and then click *Finish*. Double-clicking the *x*-axis of the resulting graph launches the *X Axis* dialog, in which we change the minimum value to 10 and the maximum value to 15, thus scaling the axis to the interval between 10 and 15 kyrs BP. Similarly, we scale the *y*-axis to the range of –30 to –50 degrees Celsius, and then select the *Numbers* tab in the *Y Axis* window, disable *Source format*, and change the *Decimal places* to zero. We can introduce numerous other modifications to the graph by, for example, adding grids from the *Insert* menu. We can now easily read from the graph that, according to the GISP2 temperature record from Alley (2000), the Younger Dryas lasted from 12.8 to 11.6 kyr BP (Fig. 3.1).

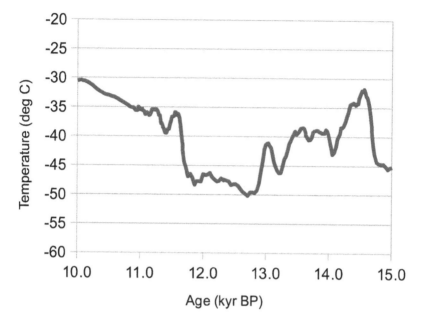

Fig. 3.1 Raw line plot of paleo-temperatures from the GISP2 ice core, created with OpenOffice Spreadsheet. Original data from the *icecore_temperature_data.txt* file from Alley (2000). The *x*-axis has been scaled to a range of 15 to 10 kyr BP, and the *y*-axis to a range of –30 to –50 degrees Celsius. We can easily see from the graph that the Younger Dryas lasted from 12.8 to 11.6 kyr BP, as indicated by an interval of consistently low temperatures.

3.6 Internet Resources: Calibrating Radiocarbon Ages

This section deals with how to access an online database to calibrate radiocarbon ages. In Section 2.4 we learned from an article by N. Roberts that the Younger Dryas cold event resulted in a drier climate in tropical East Africa. Correlating the drier episode in tropical East Africa with the high latitude Younger Dryas event requires calibration of the ice core record from Alley (2000) against the lake-level record from Roberts et al. (1993). The Younger Dryas in Greenland lasted from 12.8 to 11.6 cal. kyr BP (calibrated kiloyears before present, where the present is taken to be 1950). Figure 3 in Roberts et al. (1993) suggests that the level of Lake Magadi was low between 11.0 and 10.0 ^{14}C kyr BP (radiocarbon years before present). In order to compare the lake-level chronology from Roberts et al. (1993) with the ice core data from Alley (2000), we need to convert (calibrate) the radiocarbon ages into calibrated calendar ages (see Bradley 1999 for a detailed discussion of radiocarbon geochronologies).

Several online conversion tools are available for converting radiocarbon ages to calendar years. The most popular tool is the online version of the *CALIB* software by Stuiver and Reimer (1993) and Stuiver et al. (2005) (Fig. 3.2). The online documentation provides the required information on the various databases that are used to perform the calibration of the radiocarbon ages, depending on the environment (terrestrial or marine) from which the samples where taken. On the webpage

```
http://calib.qub.ac.uk/
```

we select *CALIB* and then access the online calibration tool by clicking on *Execute Version 6.0html*. We then get a menu with four options:

```
Data Input Menu
Calibration & Plot Options Menu
Calib Manual/Contacts
Marine Reservoir Correction Database (Lose input data)
```

We first select the *Calibration & Plot Options Menu* and enable the *1-sigma* and *2-sigma* errors, the *BP* age display, and the *Calibrated Curve*. We then select the *Data Input Menu* and type *11000* in the *Radiocarbon age BP* text field, *100* as the *Standard Deviation in age* (this being the mean 1-sigma error reported in Table 1 of Roberts et al. 1993), *Onset YD* as the *Unique*

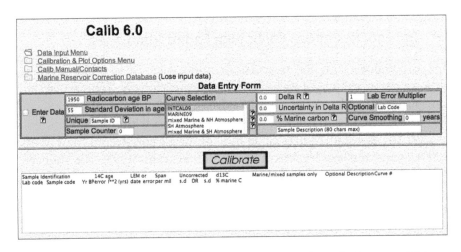

Fig. 3.2 Web interface of the online version of the *CALIB* software by Stuiver and Reimer (1993) and Stuiver et al. (2005), hosted by the 14CHRONO Centre at Queens University, Belfast. The software helps to convert (calibrate) radiocarbon ages into calibrated calendar ages (see http://calib.qub.ac.uk).

sample identifier, *INTCAL09* as the *Curve Selection* for terrestrial data, leave all other text input fields on default, and click *Enter Data*. We then change the *Unique* sample identifier to *Termination YD* and insert *10000* in the *Radiocarbon age BP* text field. Clicking *Calibrate* results in the following output:

```
                    RADIOCARBON CALIBRATION PROGRAM*
                            CALIB REV6.0.0
                Copyright 1986-2010 M Stuiver and PJ Reimer
                      *To be used in conjunction with:
           Stuiver, M., and Reimer, P.J., 1993, Radiocarbon, 35, 215-230.
                         Annotated results (text) - -
                         Export file - c14res.csv


    Onset YD
    Lab Code
    Sample Description (80 chars max)
    Radiocarbon Age BP   11000 +/-   100
    Calibration data set: intcal09.14c        # Reimer et al. 2009
    % area enclosed         cal BP age ranges       relative area under
                                                    probability
    distribution
    68.3 (1 sigma)       cal BP 12737 - 12972            0.869
                                13019 - 13061            0.131
    95.4 (2 sigma)       cal BP 12659 - 13107            1.000


    Termination YD
    Lab Code
    Sample Description (80 chars max)
    Radiocarbon Age BP   10000 +/-   100
    Calibration data set: intcal09.14c        # Reimer et al. 2009
    % area enclosed         cal BP age ranges       relative area under
                                                    probability
    distribution
    68.3 (1 sigma)       cal BP 11277 - 11628            0.942
                                11673 - 11697            0.058
    95.4 (2 sigma)       cal BP 11233 - 11828            0.959
                                11879 - 11958            0.041


    References for calibration datasets: PJ Reimer, MGL Baillie, E
    Bard, A Bayliss, JW Beck, C Bertrand, PG Blackwell, CE Buck, G
    Burr, KB Cutler, PE Damon, RL Edwards, RG Fairbanks, M Friedrich,
    TP Guilderson, KA Hughen, B Kromer, FG McCormac, S Manning, C
    Bronk Ramsey, RW Reimer, S Remmele, JR Southon, M Stuiver, S
    Talamo, FW Taylor, J van der Plicht, and CE Weyhenmeyer (2009),
    Radiocarbon 51:xxx-yyy.


    Comments: * This standard deviation (error) includes a lab error
    multiplier. ** 1 sigma = square root of (sample std. dev.^2 +
    curve std. dev.^2). ** 2 sigma = 2 x square root of (sample std.
    dev.^2 + curve std. dev.^2) where ^2 = quantity squared. [ ] =
    calibrated range impinges on end of calibration data set. 0*
    represents a "negative" age BP. 1955* or 1960* denote influence
    of nuclear testing C-14
```

```
NOTE: Cal ages and ranges are rounded to the nearest year which
may be too precise in many instances. Users are advised to round
results to the nearest 10 yr for samples with standard deviation
in the radiocarbon age greater than 50 yr.
```

The comprehensive information obtained from the online tool includes the calibrated ages for the onset of the lake-level lowstand in the Magadi basin

```
68.3 (1 sigma)       cal BP 12737 - 12972         0.869
                            13019 - 13061          0.131
```

converted from 11.0 ^{14}C kyr BP, and the termination of this lowstand

```
68.3 (1 sigma)       cal BP 11277 - 11628         0.942
                            11673 - 11697          0.058
```

converted from 10.0 ^{14}C kyr BP. The last column of the output corresponds to the mixing proportions of the two Gaussian distributions contributing to the age data. In our example, the high value of 0.899 (or 89.9 %) for the 12.737–12.972 cal. kyr BP interval for the onset of the lowstand suggests that the arid episode started at the same time as the high latitude Younger Dryas event (12.8 cal. kyr BP). The termination of the Younger Dryas event (11.6 cal. kyr BP) also falls within the termination period for the lowstand of Lake Magadi, dated at 11.277–11.628 cal. kyr BP, and suggests a synchronous climate in high and low latitudes.

3.7 Internet Resources: Insolation Data

Long-term climate fluctuations such as glacial-interglacial changes are explained by cyclic variations in the earth's orbit around the sun, know as Milankovitch cycles. These cycles affect the distribution of solar energy on our planet, causing changes in seasonality and in the heat transport between low and high latitudes, and hence in the climate. If we are running a project in tropical East Africa, we may wish to download details of these the orbital variations over the last 100 kyrs. The most popular place in which to search data of this type is again on the NCDC server of the NOAA.

```
http://ncdc.noaa.gov/
```

According to the information on their webpage, the NCDC is the world's largest active archive of weather data. Clicking on the *Paleoclimatology* icon in the lower right corner of the webpage, and then on *Climate Forcing* takes us to the directory for the orbital variations data (Fig. 3.3):

```
http://ncdc.noaa.gov/paleo/forcing.html
```

As we can see, there are three different data sets of orbital variations:

```
Orbital Variations, 5,000,000 Years, Berger and Loutre 1991
Integrated Summer Insolation, 5,000,000 Years, Huybers 2006
Orbital Variations, -50 to +20 MYrs, Laskar et al. prelim.
```

The three teams have calculated the orbital variations independently and present slightly different results. Choosing the first data set, from A. Berger and M.F. Loutre, leads us to the FTP server of the NCDC.

```
ftp://ftp.ncdc.noaa.gov/pub/data/paleo/insolation/
```

Depending which internet browser we are using, we may be required to log on as a *guest* before being directed to the text file. If your Internet browser does not launch an FTP client automatically, you can use terminal software to access the site by typing

```
ftp ftp.ncdc.noaa.gov
```

Fig. 3.3 Web interface of the *NOAA Paleoclimatology Program*, providing three data sets of orbital variations and insolation data. We use the data by Berger and Loutre (1991) as an example in this chapter (http://ncdc.noaa.gov/paleo/forcing.html).

3 | INTERNET RESOURCES FOR EARTH SCIENCE DATA

after the prompt, logging on as *anonymous*

```
Connected to ingest.ncdc.noaa.gov.
220 ProFTPD 1.3.3d Server (ProFTPD Default Installation)
[::ffff:192.168.1.20]
Name (ftp.ncdc.noaa.gov:trauth): anonymous
```

and using your email address as the password.

```
331 Anonymous login ok, send your complete email address as your
password
Password:
230 Anonymous access granted, restrictions apply
Remote system type is UNIX.
Using binary mode to transfer files.
```

After logging on, you can change the directory to

```
ftp> cd pub/data/paleo/insolation/
```

and list the directory contents by typing

```
ftp> ls
```

which yields

```
227 Entering Passive Mode (205,167,25,101,120,135)
150 Opening ASCII mode data connection for file list
-rw-rw-r--   1 147864 Oct 13  1992 bein1.dat
-rw-rw-r--   1 147864 Oct 13  1992 bein10.dat
-rw-rw-r--   1 147864 Oct 13  1992 bein11.dat
-rw-rw-r--   1 147864 Oct 13  1992 bein2.dat
-rw-rw-r--   1 147864 Oct 13  1992 bein3.dat
-rw-rw-r--   1 147864 Oct 13  1992 bein4.dat
-rw-rw-r--   1 147864 Oct 13  1992 bein5.dat
-rw-rw-r--   1 147864 Oct 13  1992 bein6.dat
-rw-rw-r--   1 147864 Oct 13  1992 bein7.dat
-rw-rw-r--   1 147864 Oct 13  1992 bein8.dat
-rw-rw-r--   1 147864 Oct 13  1992 bein9.dat
-rw-rw-r--   1   1443 Oct 13  1992 citation
-rw-rw-r--   1    762 Oct 13  1992 contents.78
-rw-rw-r--   1   1938 Oct 13  1992 contents.91
drwxrwxr-x   2   4096 Nov 14  2003 energy-balances
-rw-rw-r--   1  63169 Oct 13  1992 insol91.dec
-rw-rw-r--   1  63169 Oct 13  1992 insol91.jun
-rw-rw-r--   1 375215 Oct 13  1992 orbit91
-rw-rw-r--   1   3525 Mar 13  2008 readme_insolation.txt
226 Transfer complete
```

The directory contains a *readme* file that typically contains all relevant information on the data and the original references to be cited when using the data. We can download the file using

```
ftp> get readme_insolation.txt
```

The server's response is

```
local: readme_insolation.txt remote: readme_insolation.txt
227 Entering Passive Mode (205,167,25,101,123,254)
150 Opening BINARY mode data connection for readme_insolation.txt
(3525 bytes)
100% |*********************************|  3525       242.52
KiB/s    00:00 ETA
226 Transfer complete
3525 bytes received in 00:00 (22.42 KiB/s)
```

and the ASCII text file *readme_insolation.txt* is delivered to our hard drive.
We can open and view this file with a simple text editor. In the file, we can
read that files named *bein1.dat* to *bein11.dat* each contain 100 kyrs of orbital calculations at 1 kyr intervals. The file *orbit91* contains the variations in
the three orbital parameters eccentricity, obliquity, and precession, as well
as mid-month insolation variations, for the following latitudes and months,
65N (July), 65S (January), 15N (July) and 15S (January). The files *insol91.dec*
and *inso91.jun* contain December and June mid-month insolation variations at latitudes 90N, 60N, 30N, 0, 30S, 60S and 90S. We download the file
inso91.jun in order to be able to plot the variations in June insolation on
the 30N latitude, since this is believed to be the main driver for the African-
Asian monsoonal circulation.

```
ftp> get insol91.jun
```

The file is again delivered to our hard drive but does not have a file extension.
We change the file name from *inso91.jun* to *insol91jun.txt* to make it into a
text file, which we can then open with a text editor:

```
                              INSOLATION
       90NJune 60NJune 30NJune  0 June 30SJune 60SJune 90SJune

  0   523.30  475.95  473.93  384.08  212.28   22.76   0.00
 -1   525.28  476.99  473.77  383.21  211.13   22.09   0.00
 -2   528.91  479.55  475.14  383.58  210.69   21.50   0.00
 -3   534.08  483.54  477.99  385.18  210.95   21.01   0.00
 (cont'd)
```

The first column shows the time in kyrs, using AD 1950 as the origin and
negative numbers for the past, following the convention used for radiocarbon chronologies. All insolation values are in watts per square meter
(W/m^2). We then close the FTP connection to the NCDC server by typing

```
ftp> quit
```

We can now easily import the data into OpenOffice and follow the procedure explained in Section 3.5 to plot the data.

The second available data set was provided by P. Huybers (2006).
Clicking the corresponding link on the NCDC webpage takes us to another
webpage, which in turn provides an abstract and several links to individual
data sets. Choosing the data description again leads us to the FTP server of
the NCDC:

```
ftp://ftp.ncdc.noaa.gov/pub/data/paleo/climate_forcing/orbital_
variations/huybers2006insolation/huybers2006b.txt
```

There, the file *huybers2006b.txt* again provides all relevant information on
the data and the references to be cited when using the data. In addition to
providing numerous data files, this site also provides a MATLAB algorithm
with which to calculate the insolation values provided on the server.

The third set of data is from J. Laskar and co-workers. Clicking the corresponding link takes us to the server of the Institut de Mécanique Céleste
et de Calcul des Éphémérides in Paris, France.

```
http://imcce.fr/Equipes/ASD/insola/earth/earth.html
```

We then find a web directory by clicking on the *Source programs and
data files here* link under the heading *Solutions La2004 from -50 Myr to
+20 Myr*, which is what we are interested in.

Name	Last modified	Size
INSOLN.LA2004.BTL.100.ASC	16-Mar-2004	3.7M
INSOLN.LA2004.BTL.250.ASC	26-Apr-2004	9.0M
INSOLN.LA2004.BTL.ASC	02-Mar-2004	4.4M
INSOLP.LA2004.BTL.ASC	02-Mar-2004	1.8M
Makefile	18-Jan-2010	1.1K
README.TXT	18-Jan-2010	8.0K
insola.f	18-Jan-2010	12K
insola.par	02-Mar-2004	198
insolsub.f	18-Jan-2010	15K
prepinsol.f	02-Mar-2004	4.4K
prepsub.f	02-Mar-2004	7.5K

We again find a readme file describing the data and listing the relevant references. We select the file *INSOLN.LA2004.BTL.ASC*, which includes the
orbital parameters for the last 15 million years. According to the readme file,
the first column of the table is time in kyrs, the second to fourth columns
are the eccentricity, the obliquity (in radians) and the longitude of perihelion
from moving equinox (in radians). We again rename the file as *INSOLN.
LA2004BTLASC.txt* and import it into OpenOffice in order to plot the data.

3.8 Internet Resources: Tephrabase

The last example of Internet resources for geoscientific data is *Tephrabase*, which is hosted by the *School of GeoSciences* of the *University of Edinburgh*, United Kingdom (Fig. 3.4):

```
http://tephrabase.org/
```

According to their webpage, Tephrabase was originally designed for use by scientists involved in tephrochronological research in north-west Europe. Tephrochronology uses volcanic ash layers as unique time markers within marine and terrestrial sedimentary sections. The ash layers can be either radiometrically dated, or correlated geochemically with other layers that have already been dated. Searches on Tephrabase are divided into tephra from either Iceland and north-west Europe, or central Mexico. As an example, we can use Tephrabase to identify an as yet undated ash layer that we found in a marine sediment core from the North Atlantic. The oxygen isotope chronology of the core indicates an age for the ash is within the range of 10–12 ^{14}C kyrs BP. This implies that the associated volcanic eruption occurred either during, or close to, Younger Dryas times.

We select *Introduction* from the *Searches* menu bar on the left, click on *Dates* in the *Iceland and north-west Europe Searches* section, and then on *Radiocarbon* in order to perform a search based on ^{14}C dates. Please note that the dates provided on this webpage are radiocarbon ages, and not

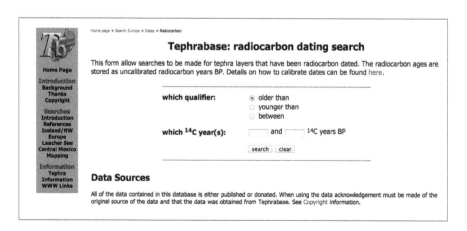

Fig. 3.4 The *Tephrabase* web interface, hosted by the School of GeoSciences of the University of Edinburgh, United Kingdom. The database helps with the identification of ash layers within the time range of 10-12 radiocarbon years before present.

not calendar ages (see Section 3.7). We select *between* and enter *10000* and *12000* ^{14}C kyrs BP, without thousand separators. The resulting page suggests the *Vedde Tephra* as the only possible ash layer within this time interval. Clicking on *Vedde Tephra* yields several locations in Sweden where the ash layer has been identified, geochemically analyzed, and dated at around 10,300 ± 100 ^{14}C kyrs BP. Indeed, the Vedda Ash is an important time marker in deep-sea cores from the North Atlantic.

3.9 Organizing Data in a Computer

During the course of a project we accumulate large volumes of diverse data on the hard drives of our computers. In addition to a variety of information relating to the project, such as the addresses of collaborators in an address book, the project schedule in a calendar, and discussion threads with colleagues in email software, we also collect field photographs, measurements on samples, laboratory analyses, processed data, and published results. The rising flood of data acquired by the project requires well-organized data file directories on the computer.

We will, however, not discuss in this section specific software for data management or methods for analyzing large data sets. We will simply put together a few thoughts on how to store and organize the data in a computer using the on-board tools of the operating system. A filing system, of course, has to be adaptable to a growing project. As an example, we set up a top-level directory called *Younger Dryas East Africa*. We then create various levels of subfolders for various types of data, using the sorting function of the operating system to establish a hierarchy in our project directory, e.g., by labeling the subfolders 01, 02, 03 ..., or by using years such as 2008, 2009, 2010, 2011 ... to sort the subfolders.

```
Project Younger Dryas East Africa
    01 Research proposal and permits
        011 Outlines and ideas
        012 Materials
        013 Proposal text
        014 Proposal figures
        015 Correspondence
    02 Data from other people
    03 My field observations
        031 Field maps
        032 Field photos
        033 Field measurements
            0331 Method 1
            0332 Method 2
            ...
        034 Field samples
```

```
04 My laboratory observations
    041 Laboratory photos
    042 Laboratory measurements
        0431 Method 1
        0432 Method 2
        . . .
    043 Laboratory samples and subsamples
    044 Laboratory processed samples
05 My processed field and laboratory data
    051 Processing method 1
    052 Processing method 2
        . . .
06 Literature
07 Correspondence
08 Logistics
09 Publications
    091 Conference Presentations
    092 Manuscripts
        2008 Paper Younger Dryas East Africa
        2009 Paper Lake Magadi Lake Sediments
        2009 Paper Mag Sus Record Lake Magadi
            . . .
    093 Reports
        . . .
10 Other
```

When storing data on a computer, filenames should not have any special characters or punctuation, other than full stop preceding the file extension. File extensions help to assign a file to a particular software. The most popular file extensions and their file types are

```
*.doc, *.xls, *.ppt  Microsoft Word, Excel and PowerPoint
*.txt, *.rtf         ASCII text and Microsoft Rich Text
*.htm, *.html        HyperText
*.exe, *.app         Executables under Windows and Mac OS X
*.zip, *.tar         Compressed files
*.gif, *.jpg, *.bmp  Various raster graphics
*.ps, *.eps, *.pdf   Various vector graphics
*.ai, *.psd, *.indd  Adobe Illustrator, Photoshop and InDesign
*.m, *.mat           MATLAB text and binary formats
```

Most of these file types have been described in the preceding text, or will be used in examples presented in the following chapters. The more specific file types include control characters that are not visible to the user but provide the relevant information concerning the formats of text and graphics, to the software. Most software products include converters for interpreting file types from other software products. In many examples, however, the appropriate file types for exchanging information are ASCII for text, TIFF or JPEG for pixel graphics, and PS, EPS or PDF for vector graphics. Other types of files will be explained in Chapters 5, 6 and 7.

Meaningful file names should be used (e.g., *odp659_oxygenisotopes.xls* or *BG08_14C_Nov08.pdf*) containing information on the type of data stored in the file, and including, for example, the name of a marine sediment core or sampling location, the method used to generate the data, and the date on which the file was created. Different versions of your files, such as processed data and manuscripts, should be retained in case you ever need to revert to a previous version of your work. If you are working on the same file as your collaborators, make sure that when your colleagues return the file they add their initials to the file name on the version that they have worked on, for example:

```
paper_youngerdryaseastafrica_vs1.doc
paper_youngerdryaseastafrica_vs2.doc
paper_youngerdryaseastafrica_vs2_MHT.doc
paper_youngerdryaseastafrica_vs3.doc
...
figure1_youngerdryaseastafrica_vs1.ai
figure1_youngerdryaseastafrica_vs2.ai
figure2_youngerdryaseastafrica_vs1.ai
figure2_youngerdryaseastafrica_vs1_MHT.ai
figure2_youngerdryaseastafrica_vs1_ES.ai
figure3_youngerdryaseastafrica_vs1.ai
...
```

Finally, be sure to back up your data regularly. Submissions of doctoral theses have in the past been delayed, either because a computer hard drive crashed a few days before the deadline, or because the entire computer system was stolen. Doctoral candidates are typically remiss about backing up during the last few weeks before submission and fully automated differential backups on a separated hard drive are therefore recommended. Examples of software that can be used to perform differential backups are *Apple Time Machine*, *Microsoft Windows Backup*, and the third-party *Genie-Soft Backup Manager*. Ideally, two backup hard drives should be used alternately on a daily basis and stored in different locations. In addition to these hard drive backup systems, there are a number of online backup services available, such as *Dropbox* (http://dropbox.com) and *Apple iCloud* (http://apple.com/icloud).

Recommended Reading

Berger A, Loutre MF (1991) Insolation values for the climate of the last 10 million years. Quaternary Sciences Review 10:297–317, 1991

Bradley RS (1999) Paleoclimatology – Second Edition: Reconstructing Climates of the Quaternary. Academic Press, Burlington

Huybers P (2006) Early Pleistocene Glacial Cycles and the Integrated Summer Insolation Forcing. Science 313:508–511

Laskar J, Robutel P, Joutel F, Gastineau M, Correia ACM, Levrard B (2004) A long-term numerical solution for the insolation quantities of the Earth. Astronomy and Astrophysics 428:261

Laskar J, Joutel F, Boudin F (1993) Orbital, precessional and insolation quantities for the Earth from –20 Myr to +10 Myr. Astronomy and Astrophysics 270: 522

Newton AJ, Gittings B, Stuart N (1997) Designing a scientific database Query Server using the World Wide Web: The example of Tephrabase. Innovations in GIS 4, Taylor and Francis, London, 251–266

Newton AJ (1996) Tephrabase. A Tephrochronological Database. Quaternary Newsletter 78:8–13

Ruddiman WF (2000) Earth's Climate – Past and Future. W.H. Freeman and Company, New York

Stuiver M, Reimer PJ (1993) Extended ^{14}C database and revised CALIB radiocarbon calibration program Radiocarbon 35:215–230

Stuiver M, Reimer PJ, Reimer RW (2005) CALIB 5.0 – WWW program and documentation (http://calib.qub.ac.uk)

4 MATLAB as a Visualization Tool

4.1 Introduction

MATLAB® is a software package developed by *The MathWorks Inc.*, founded by Cleve Moler, Jack Little and Steve Bangert in 1984, which has its headquarters in Natick, Massachusetts (http://mathworks.com). MATLAB was designed to perform mathematical calculations, to analyze and visualize data, and to facilitate the writing of new software programs. The advantage of this software is that it combines comprehensive math and graphics functions with a powerful high-level language. Since MATLAB contains a large library of ready-to-use routines for a wide range of applications, the user can solve technical computing problems much more quickly than with traditional programming languages, such as C++ and FORTRAN. The standard library of functions can be significantly expanded by add-on toolboxes, which are collections of functions for special purposes such as image processing, creating map displays, performing geospatial data analysis or solving partial differential equations.

During the last few years MATLAB has become an increasingly popular tool in earth sciences. It has been used for finite element modeling, processing of seismic data, analyzing satellite imagery, and for the generation of digital elevation models from satellite data. The continuing popularity of the software is apparent in published scientific literature, and many conference presentations have also made reference to MATLAB. Universities and research institutions have recognized the need for MATLAB training for staff and students, and many earth science departments across the world now offer MATLAB courses for undergraduates. *The MathWorks Inc.* provides classroom kits for teachers at a reasonable price, and it is also possible for students to purchase a low-cost edition of the software. This student version provides an inexpensive way for students to improve their MATLAB skills.

The following sections contain a tutorial-style introduction to MATLAB, to the setup on the computer (Section 4.2), the syntax (Section 4.3), data

M. H. Trauth and E. Sillmann, *MATLAB® and Design Recipes for Earth Sciences*, DOI: 10.1007/978-3-642-32544-1_4, © Springer-Verlag Berlin Heidelberg 2013

input and output (Sections 4.4 and 4.5), programming (Section 4.6), and visualization (Section 4.7). Advanced sections are also included on generating M-files to regenerate graphs (Section 4.8) and on publishing M-files (Section 4.9). The reader is recommended to go through the entire chapter in order to obtain a good knowledge of the software before proceeding to the following chapter. A more detailed introduction can be found in the *MATLAB Getting Started Guide* (The MathWorks 2011) which is available in print form, online and as PDF file.

4.2   Getting Started with MATLAB

The software package comes with extensive documentation, tutorials and examples. The first three chapters of the book *MATLAB Getting Started Guide* (The MathWorks 2010) are directed at beginners. The chapters on programming, creating graphical user interfaces (GUIs) and development environments are aimed at more advanced users. Since *MATLAB Getting Started Guide* provides all the information required to use the software, this introduction concentrates on the most relevant software components and tools used in the following chapters of this book.

After the installation of MATLAB, the software is launched either by clicking the shortcut icon on the desktop or by typing

```
matlab
```

in the operating system prompt. The software then comes up with several window panels (Fig. 4.1). The default desktop layout includes the *Current Folder* panel that lists the files in the directory currently being used. The *Command Window* presents the interface between the software and the user, i.e., it accepts MATLAB commands typed after the prompt, >>. The *Workspace* panel lists the variables in the MATLAB workspace, which is empty when starting a new software session. The *Command History* panel records all operations previously typed into the Command Window and enables them to be recalled by the user. In this book we mainly use the Command Window and the built-in *Editor*, which can be launched by typing

```
edit
```

or by selecting the *Editor* from the *Desktop* menu. By default, the software stores all of your MATLAB-related files in the startup folder named *MATLAB*. Alternatively, you can create a personal working directory in

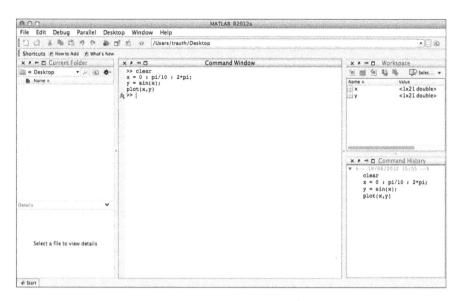

Fig. 4.1 Screenshot of the MATLAB default desktop layout including the *Current Folder* (left in the figure), the *Command Window* (center), the *Workspace* (upper right) and *Command History* (lower right) panels. This book uses only the Command Window and the built-in *Editor*, which can be called up by typing edit after the prompt. All information provided by the other panels can also be accessed through the Command Window.

which to store your MATLAB-related files. You should then make this new directory the working directory using the *Current Folder* panel or the *Folder Browser* at the top of the MATLAB desktop. The software uses a *search path* to find MATLAB-related files, which are organized in directories on the hard disk. The default search path includes only the *MATLAB R2011b* directory (or a similar folder if any other release of the software is being used) that has been created by the installer in the applications folder, and the default working directory named *MATLAB*. To see which directories are in the search path or to add new directories, select *Set Path* from the *File* menu, and use the *Set Path* dialog box. The modified search path is saved in a file *pathdef.m* on your hard disk. The software will then in future read the contents of this file and direct MATLAB to use your custom path list.

4.3 The Syntax of MATLAB

The name MATLAB stands for *matrix laboratory*. The classic object handled by MATLAB is a *matrix*, i.e., a rectangular two-dimensional *array* of numbers. A simple 1-by-1 matrix or array is a scalar. Matrices with one

MATLAB AS A VISUALIZATION TOOL

4

column or row are vectors, time series or other one-dimensional data fields. An *m*-by-*n* matrix or array can be used for a digital elevation model or a grayscale image. Red, green and blue (RGB) color images are usually stored as three-dimensional arrays, i.e., the colors red, green and blue are represented by an *m*-by-*n*-by-3 array.

Before proceeding, we need to clear the workspace by typing

```
clear
```

after the prompt in the Command Window. Clearing the workspace is always recommended before working on a new MATLAB project. Entering matrices or arrays in MATLAB is easy. To enter an arbitrary matrix, type

```
A = [2 4 3 7; 9 3 -1 2; 1 9 3 7; 6 6 3 -2]
```

which first defines a variable A, then lists the elements of the matrix in square brackets. The rows of A are separated by semicolons, whereas the elements of a row are separated by blank spaces, or alternatively, by commas. After pressing *return*, MATLAB displays the matrix

```
A =
     2     4     3     7
     9     3    -1     2
     1     9     3     7
     6     6     3    -2
```

Displaying the elements of A could be problematic for very large matrices, such as digital elevation models consisting of thousands or millions of elements. To suppress the display of a matrix or the result of an operation in general, the line should be ended with a semicolon.

```
A = [2 4 3 7; 9 3 -1 2; 1 9 3 7; 6 6 3 -2];
```

The matrix A is now stored in the workspace and we can carry out some basic operations with it, such as computing the sum of elements,

```
sum(A)
```

which results in the display

```
ans =
    18    22     8    14
```

Since we did not specify an output variable, such as A for the matrix entered above, MATLAB uses a default variable ans, short for *answer* or *most recent answer*, to store the results of the calculation. In general, we should define variables since the next computation without a new variable name will overwrite the contents of ans.

The above example illustrates an important point about MATLAB: the software prefers to work with the columns of matrices. The four results of sum(A) are obviously the sums of the elements in each of the four columns of A. To sum all elements of A and store the result in a scalar b, we simply need to type

```
b = sum(sum(A));
```

which first sums the columns of the matrix and then the elements of the resulting vector. We now have two variables, A and b, stored in the workspace. We can easily check this by typing

```
whos
```

which is one the most frequently-used MATLAB commands. The software then lists all variables in the workspace, together with information about their sizes or dimensions, number of bytes, classes and attributes (see Section 4.5 for details about classes and attributes of objects).

```
Name      Size           Bytes  Class     Attributes
A         4x4              128  double
ans       1x4               32  double
b         1x1                8  double
```

Note that by default MATLAB is case sensitive, i.e., A and a can define two different variables. In this context, it is recommended that capital letters be used for matrices and lower-case letters for vectors and scalars. We could now delete the contents of the variable ans by typing

```
clear ans
```

Next, we will learn how specific matrix elements can be accessed or exchanged. Typing

```
A(3,2)
```

simply yields the matrix element located in the third row and second column, which is 9. The matrix indexing therefore follows the rule *(row, column)*. We can use this to replace single or multiple matrix elements. As an example, we type

```
A(3,2) = 30
```

to replace the element A(3,2) by 30 and to display the entire matrix.

```
A =
     2      4      3      7
     9      3     -1      2
     1     30      3      7
     6      6      3     -2
```

If we wish to replace several elements at one time, we can use the *colon operator*. Typing

```
A(3,1:4) = [1 3 3 5]
```

or

```
A(3,:) = [1 3 3 5]
```

replaces all elements of the third row of the matrix A. The colon operator also has several other uses in MATLAB, for instance as a shortcut for entering matrix elements such as

```
c = 0 : 10
```

which creates a row vector containing all integers from 0 to 10. The resultant MATLAB response is

```
c =
     0    1    2    3    4    5    6    7    8    9   10
```

Note that this statement creates 11 elements, i.e., the integers from 1 to 10 and the zero. A common error when indexing matrices is to ignore the zero and therefore expect 10 elements instead of 11 in our example. We can check this from the output of whos.

```
Name        Size              Bytes  Class       Attributes
A           4x4                 128  double
ans         1x1                   8  double
b           1x1                   8  double
c           1x11                 88  double
```

The above command creates only integers, i.e., the interval between the vector elements is one unit. However, an arbitrary interval can be defined, for example 0.5 units. This is later used to create evenly-spaced time axes for time series analysis. Typing

```
c = 1 : 0.5 : 10
```

results in the display

```
c =
   Columns 1 through 6
```

```
   1.0000     1.5000     2.0000     2.5000     3.0000     3.5000
Columns 7 through 12
   4.0000     4.5000     5.0000     5.5000     6.0000     6.5000
Columns 13 through 18
   7.0000     7.5000     8.0000     8.5000     9.0000     9.5000
Column 19
  10.0000
```

which autowraps the lines that are longer than the width of the Command Window. The display of the values of a variable can be interrupted by pressing *Ctrl+C* (*Control+C*) on the keyboard. This interruption affects only the output in the Command Window, whereas the actual command is processed before displaying the result.

MATLAB provides standard arithmetic operators for addition, +, and subtraction, -. The asterisk, *, denotes matrix multiplication involving inner products between rows and columns. As an example, we multiply the matrix A with a new matrix B

```
B = [4 2 6 5; 7 8 5 6; 2 1 -8 -9; 3 1 2 3];
```

the matrix multiplication is then

```
C = A * B'
```

where ' is the complex conjugate transpose, which turns rows into columns and columns into rows. This generates the output

```
C =
    69    103    -79     37
    46     94     11     34
    75    136    -76     39
    44     93     12     24
```

In linear algebra, matrices are used to keep track of the coefficients of linear transformations. The multiplication of two matrices represents the combination of two linear transformations into a single transformation. Matrix multiplication is not commutative, i.e., A*B' and B*A' yield different results in most cases. Similarly, MATLAB allows matrix divisions, right / and left \ representing different transformations. Finally, the software also allows powers of matrices ^.

In earth sciences, however, matrices are often simply used as two-dimensional arrays of numerical data rather than an array representing a linear transformation. Arithmetic operations on such arrays are carried out element-by-element. While this does not make any difference in addition and subtraction, it does affect multiplicative operations. MATLAB uses a dot as part of the notation for these operations.

4 | MATLAB AS A VISUALIZATION TOOL

As an example, multiplying A and B element-by-element is performed by typing

```
C = A .* B
```

which generates the output

```
C =
      8      8     18     35
     63     24     -5     12
      2      3    -24    -45
     18      6      6     -6
```

4.4 Data Storage and Handling

This section deals with how to store, import, and export data with MATLAB. Many of the data formats typically used in earth sciences have to be converted before being analyzed with MATLAB. Alternatively, the software provides several import routines to read many binary data formats in earth sciences, such as those used to store digital elevation models and satellite data.

The simplest way to exchange data between a certain piece of software and MATLAB is using the ASCII format. Although the newer versions of MATLAB provide various import routines for file types such as Microsoft Excel binaries, most data arrive in the form of ASCII files. Consider a simple data set stored in a table such as

```
SampleID       Percent C      Percent S
101            0.3657         0.0636
102            0.2208         0.1135
103            0.5353         0.5191
104            0.5009         0.5216
105            0.5415         -999
106            0.501          -999
```

The first row contains the names of the variables and the columns provide the data for each sample. The absurd value -999 indicates missing data in the data set. Two things have to be changed to convert this table into MATLAB format. First, MATLAB uses NaN as the representation for *Not-a-Number* that can be used to mark missing data or gaps. Second, a percent sign, %, should be added at the beginning of the first line. The percent sign is used to indicate nonexecutable text within the body of a program. This text is normally used to include comments in the code.

```
%SampleID      Percent C      Percent S
101            0.3657         0.0636
```

```
102              0.2208          0.1135
103              0.5353          0.5191
104              0.5009          0.5216
105              0.5415          NaN
106              0.501           NaN
```

MATLAB will ignore any text appearing after the percent sign and continue processing on the next line. After editing this table in a text editor, such as the *MATLAB Editor*, it can be saved as ASCII text file *geochem_data.txt* in the current working directory (Fig. 4.2). The MATLAB workspace should first be cleared by typing

```
clear
```

after the prompt in the Command Window. MATLAB can now import the data from this file with the `load` command.

```
load geochem_data.txt
```

MATLAB then loads the contents of the file and assigns the matrix to a vari-

Fig. 4.2 Screenshot of MATLAB *Editor* showing the content of the file *geochem_data.txt*. The first line of the text is commented by a percent sign at the beginning of the line, followed by the actual data matrix.

able `geochem_data` specified by the filename *geochem_data.txt*. Typing

```
whos
```

yields

```
Name                    Size             Bytes  Class      Attributes
geochem_data            6x3                144  double
```

The command `save` now allows workspace variables to be stored in a binary format.

```
save geochem_data_new.mat
```

MAT-files are double precision binary files using *.mat* as extension. The advantage of these binary MAT-files is that they are independent of the computer platforms running different floating-point formats. The command

```
save geochem_data_new.mat geochem_data
```

saves only the variable `geochem_data` instead of the entire workspace. The option `-ascii`, for example

```
save geochem_data_new.txt geochem_data -ascii
```

again saves the variable `geochem_data`, but in an ASCII file named *geochem_data_new.txt*. In contrast to the binary file *geochem_data_new.mat*, this ASCII file can be viewed and edited using the MATLAB Editor or any other text editor.

4.5   Data Structures and Classes of Objects

The default data type or *class* in MATLAB is *double precision* or `double`, which stores data in a 64-bit array. This double precision array allows storage of the sign of a number (first bit), the exponent (bits 2 to 12) and roughly 16 significant decimal digits between approximately 10^{-308} and 10^{+308} (bits 13 to 64). As an example, typing

```
clear

rand('seed',0)
A = rand(3,4)
```

creates a 3-by-4 array of random numbers with double precision. We use the function `rand` that generates uniformly distributed pseudorandom numbers within the open interval (0,1). To obtain identical data values, we reset the random number generator by using the integer 0 as *seed*. Since we did not use a semicolon here we get the output

```
A =
    0.2190    0.6793    0.5194    0.0535
    0.0470    0.9347    0.8310    0.5297
    0.6789    0.3835    0.0346    0.6711
```

By default, the output is in a scaled fixed point format with 5 digits, e.g., 0.2190 for the (1,1) element of A. Typing

```
format long
```

switches to a fixed point format with 16 digits for double precision. Recalling A by typing

```
A
```

yields the output

```
A =
  Columns 1 through 3
    0.218959186328090    0.679296405836612    0.519416372067955
    0.047044616214486    0.934692895940828    0.830965346112365
    0.678864716868319    0.383502077489859    0.034572110527461
  Column 4
    0.053461635044525
    0.529700193335163
    0.671149384077242
```

which autowraps those lines that are longer than the width of the Command Window. The command format does not affect how the computations are carried out, i.e., the precision of the computation results is not changed. The precision is, however, affected by converting the data type from *double* to 32-bit *single precision*. Typing

```
B = single(A)
```

yields

```
B =
    0.2189592    0.6792964    0.5194164    0.0534616
    0.0470446    0.9346929    0.8309653    0.5297002
    0.6788647    0.3835021    0.0345721    0.6711494
```

Although we have switched to format long, only 8 digits are displayed. The command whos lists the variables A and B with information on their sizes or dimensions, number of bytes and classes:

```
Name        Size               Bytes  Class      Attributes
A           3x4                   96  double
B           3x4                   48  single
```

The default class `double` is used in all MATLAB operations in applications where the physical memory of the computer is not a limiting factor, whereas `single` is used when working with large data sets. The double precision variable A, whose size is 3×4 elements, requires $3 \times 4 \times 64 = 768$ bits or $768/8 = 96$ bytes of memory, whereas B requires only 48 bytes and so has half the memory requirement of A. Introducing at least one complex number to A doubles the memory requirement since both real and imaginary parts are double precision by default. Switching back to `format short` and typing

```
format short
A(1,3) = 4i + 3
```

yields

```
A =
    0.2190          0.6793          3.0000 + 4.0000i    0.0535
    0.0470          0.9347          0.8310              0.5297
    0.6789          0.3835          0.0346              0.6711
```

and the variable listing is now

```
Name        Size               Bytes  Class      Attributes
A           3x4                  192  double     complex
B           3x4                   48  single
```

indicating the class `double` and the attribute `complex`.

MATLAB also works with even smaller data types such as 1-bit, 8-bit and 24-bit data in order to save memory. These data types are used to store digital elevation models or images. For example, m-by-n pixel RGB true color images are usually stored as three-dimensional arrays, i.e., the three colors are represented by an m-by-n-by-3 array. Such multi-dimensional arrays can be generated by concatenating three two-dimensional arrays representing the m-by-n pixels of an image. First, we generate a 100-by-100 array of uniformly distributed random numbers in the range of 0 to 1. We then multiply the random numbers by 256 and round the results towards plus infinity using the function `ceil` to get values between 1 and 256.

```
clear

I1 = 256 * rand(100,100); I1 = ceil(I1);
I2 = 256 * rand(100,100); I2 = ceil(I2);
I3 = 256 * rand(100,100); I3 = ceil(I3);
```

The command `cat` concatenates the three two-dimensional arrays (8 bits each) to a three-dimensional array (3×8 bits = 24 bits).

```
I = cat(3,I1,I2,I3);
```

Since RGB images are represented by integer values between 1 and 256 for each color, we convert the 64-bit double precision values to unsigned 8-bit integers using `uint8`.

```
I = uint8(I);
```

Typing `whos` then yields

```
Name          Size              Bytes  Class      Attributes
I             100x100x3         30000  uint8
I1            100x100           80000  double
I2            100x100           80000  double
I3            100x100           80000  double
```

Since 8 bits can be used to store 256 different values, this data type can be used to store integer values between 1 and 256, whereas using `int8` to create signed 8-bit integers generates values between –128 and +127. The value of zero requires one bit and therefore there is no space to store +128. Finally, `imshow` can be used to display the three-dimensional array as a true color image.

```
imshow(I)
```

We next introduce *structure arrays* as a MATLAB data type. Structure arrays are multi-dimensional arrays with elements accessed by textual field designators. These arrays are data containers that are particularly helpful in storing any kind of information about a sample in a single variable. As an example, we can generate a structure array `sample_1` that includes the image array `I` defined in the previous example as well as other types of information about a sample, such as the name of the sampling location, the date of sampling, and geochemical measurements, stored in a 10-by-10 array.

```
sample_1.location = 'Plougasnou';
sample_1.date = date;
sample_1.image = I;
sample_1.geochemistry = rand(10,10);
```

The first layer of the structure array `sample_1` contains a character array, i.e., a two-dimensional array of the data type `char` containing a character string. We can create such an array by typing

MATLAB AS A VISUALIZATION TOOL

4

```
location = 'Plougasnou';
```

We can list the size, class and attributes of a single variable such as `loca-tion` by typing

```
whos location
```

and learn from

```
Name         Size            Bytes  Class     Attributes
location     1x10               20  char
```

that the size of this character array `location` corresponds to the number of characters in the word *Plougasnou*. Character arrays are 16 bit arrays, i.e., $2^{16} = 65{,}536$ different characters can be stored in such arrays. The character string `location` therefore requires $10 \times 16 = 160$ bits or $160/8 = 20$ bytes of memory. Also the second layer `datum` in the structure array `sample_1` contains a character string generated by `date` that yields a string containing the current date in `dd-mm-yyyy` format. We access this particular layer in `sample_1` by typing

```
sample_1.date
```

which yields

```
ans =
   06-Oct-2009
```

The third layer of `sample_1` contains the image created in the previous example, whereas the fourth layer contains a 10-by-10 array of uniformly-distributed pseudorandom numbers. All layers of `sample_1` can be listed by typing

```
sample_1
```

resulting in the output

```
sample_1 =
        location: 'Plougasnou'
            date: '06-Oct-2009'
           image: [100x100x3 uint8]
     geochemistry: [10x10 double]
```

This represents a list of the layers `location`, `date`, `image` and `geochemistry` within the structure array `sample_1`. Some variables are listed in full, whereas larger data arrays are only represented by their size. In the list of the layers within the structure array `sample_1`, the array `image` is characterized by its size `100x100x3` and the class `uint8`. The variable `geochemistry` in the last layer of the structure array contains a 10-by-10 array of double precision numbers. The command

```
whos sample_1
```

does not list the layers in `sample_1` but the name of the variable, the bytes and the class `struct` of the variable.

```
Name            Size              Bytes  Class    Attributes
sample_1        1x1               31546  struct
```

MATLAB also has *cell arrays* as an alternative to structure arrays. Both classes or data types are very similar and are containers of different types and sizes of data. The most important difference between the two is that the containers of a structure array are *named fields*, whereas a cell array uses *numerically indexed cells*. Structures are often used in applications where organization of the data is of high importance. Cell arrays are often used when working with data that is intended for processing by index in a programming control loop.

4.6 Scripts and Functions

MATLAB is a powerful programming language. All files containing MATLAB code use *.m* as an extension and are therefore called *M-files*. These files contain ASCII text and can be edited using a standard text editor. However, the built-in Editor color-highlights various syntax elements such as comments in green, keywords such as *if, for* and *end* in blue and character strings in pink. This syntax highlighting facilitates MATLAB coding.

MATLAB uses two types of M-files: *scripts* and *functions*. Whereas scripts are a series of commands that operate on data in the workspace, functions are true algorithms with input and output variables. The advantages and disadvantages of both types of M-file will now be illustrated by an example. We first start the Editor by typing

```
edit
```

This opens a new window named *untitled*. Next, we generate a simple MATLAB script by typing a series of commands to calculate the average of the elements of a data vector x.

```
[m,n] = size(x);
if m == 1
   m = n;
end
```

```
sum(x)/m
```

The first line of the *if loop* yields the dimensions of the variable x using the command `size`. In our example, x should be either a column vector with dimensions `(m,1)` or a row vector with dimensions `(1,n)`. The `if` statement evaluates a logical expression and executes a group of commands if this expression is true. The `end` keyword terminates the last group of commands. In the example, the `if` loop picks either m or n depending on whether m==1 is false or true. Here, the double equal sign `'=='` makes element by element comparisons between the variables (or numbers) to the left and right of the equal signs and returns a matrix of the same size, made up of elements set to logical 1 where the relation is true and elements set to logical 0 where it is not. In our example, `m==1` returns 1 if m equals 1 and 0 if m equals any other value. The last line of the `if` loop computes the average by dividing the sum of elements by m or n. We do not use a semicolon here in order to allow the output of the result. We can now save our new M-file as *average.m* and type

```
clear

x = [3 6 2 -3 8];
```

in the Command Window to define an example vector x. We then type

```
average
```

after the Command Window's prompt, without the extension .*m*, to run our script and obtain the average of the elements of the vector x as output.

```
ans =
    3.2000
```

After typing

```
whos
```

we see that the workspace now contains

```
Name        Size          Bytes  Class      Attributes
ans         1x1               8  double
m           1x1               8  double
n           1x1               8  double
x           1x5              40  double
```

The listed variables are the example vector x and the output of the function `size`, m and n. The result of the operation is stored in the variable ans.

Since the default variable ans might be overwritten during one of the succeeding operations, we need to define a different variable. Typing

```
a = average
```

however, results in the error message

```
??? Attempt to execute SCRIPT average as a function.
```

Obviously, we cannot assign a variable to the output of a script. Moreover, all variables defined and used in the script appear in the workspace; in our example these are the variables m and n. Scripts contain sequences of commands that are applied to variables in the workspace. MATLAB functions, however, allow inputs and outputs to be defined. They do not automatically import variables from the workspace. To convert the above script into a function, we have to introduce the following modifications (Fig. 4.3):

```
function y = average(x)
%AVERAGE      Average value.
%      AVERAGE(X) is the average of the elements in the vector X.

% By Martin Trauth, Oct 6, 2009

[m,n] = size(x);
if m == 1
    m = n;
end
y = sum(x)/m;
```

The first line now contains the keyword function, the function name average, the input x and the output y. The next two lines contain comments as indicated by the percent sign, separated by an empty line. The second comment line contains the author's name and the version of the M-file. The rest of the file contains the actual operations. The last line now defines the value of the output variable y, and this line is terminated by a semicolon to suppress the display of the result in the Command Window. Next we type

```
help average
```

which displays the first block of contiguous comment lines. The first executable statement (or blank line in our example) effectively ends the help section and therefore the output of *help*. Now we are independent of the variable names used in our function. The workspace can now be cleared and a new data vector defined.

```
clear
```

```
data = [3 6 2 -3 8];
```

Our function can then be run by the statement

```
result = average(data);
```

This clearly illustrates the advantages of functions compared to scripts. Typing

```
whos
```

results in

```
Name          Size              Bytes  Class      Attributes
data          1x5                  40  double
result        1x1                   8  double
```

revealing that all variables used in the function do not appear in the work-space. Only the input and output, as defined by the user, are stored in the

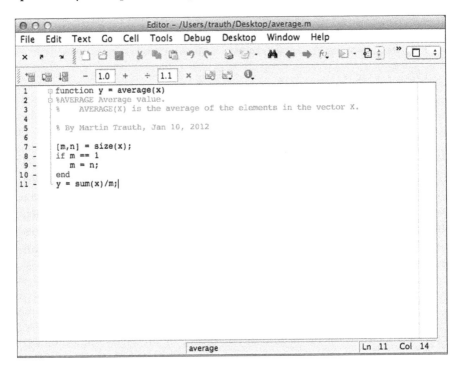

Fig. 4.3 Screenshot of the MATLAB *Editor* showing the function `average`. The function starts with a line containing the keyword `function`, the name of the function `average`, the input variable `x` and the output variable `y`. The subsequent lines contain the output for `help average`, the copyright and version information, and also the actual MATLAB code for computing the average using this function.

workspace. The M-files can therefore be applied to data as if they were real functions, whereas scripts contain sequences of commands that are applied to the variables in the workspace.

4.7 Basic Visualization Tools

MATLAB provides numerous routines for displaying data as graphs. This section introduces the most important graphics functions. The graphs can be modified, printed and exported to be edited with graphics software other than MATLAB. The simplest function producing a graph of a variable y versus another variable x is `plot`. First, we define two vectors x and y, where y is the sine of x. The vector x contains values between 0 and 2π with $\pi/10$ increments, whereas y is the element-by-element sine of x.

```
clear

x = 0 : pi/10 : 2*pi;
y = sin(x);
```

These two commands result in two vectors with 21 elements each, i.e., two 1-by-21 arrays. Since the two vectors x and y have the same length, we can use `plot` to produce a linear 2D graph y against x.

```
plot(x,y)
```

This command opens a *Figure Window* named *Figure 1* with a gray background, an *x*-axis ranging from 0 to 7, a *y*-axis ranging from –1 to +1 and a blue line. We may wish to plot two different curves in a single plot, for example the sine and the cosine of x in different colors. The command

```
x = 0 : pi/10 : 2*pi;
y1 = sin(x);
y2 = cos(x);

plot(x,y1,'r--',x,y2,'b-')
```

creates a dashed red line displaying the sine of x and a solid blue line representing the cosine of this vector (Fig. 4.4). If we create another plot, the window *Figure 1* will be cleared and a new graph displayed. The command `figure`, however, can be used to create a new figure object in a new window.

```
plot(x,y1,'r--')
figure
plot(x,y2,'b-')
```

Instead of plotting both lines in one graph simultaneously, we can also plot
the sine wave, hold the graph and then plot the second curve. The command
hold is particularly important for displaying data while using different plot
functions, for example if we wish to display the second graph as a bar plot.

```
plot(x,y1,'r--')
hold on
bar(x,y2)
hold off
```

This command plots y1 versus x as dashed line, whereas y2 versus x is
shown as a group of blue vertical bars. Alternatively, we can plot both graphs
in the same Figure Window but in different plots using subplot. The syntax subplot(m,n,p) divides the Figure Window into an *m*-by-*n* matrix

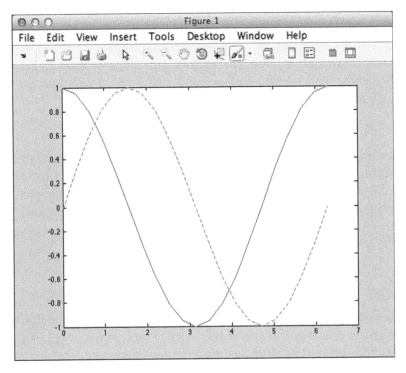

Fig. 4.4 Screenshot of the MATLAB *Figure Window* showing two curves in different
colors and line types. The Figure Window allows editing of all elements of the graph after
selecting *Edit Plot* from the *Tools* menu. Double clicking on the graphics elements opens an
options window for modifying the appearance of the graphs. The graphics can be exported
using *Save as* from the *File* menu. The command *Generate M-File* from the *File* menu
creates MATLAB code from an edited graph.

of display regions and makes the pth display region active.

```
subplot(2,1,1), plot(x,y1,'r--')
subplot(2,1,2), bar(x,y2)
```

For example, the Figure Window is divided into two rows and one column. The 2D linear plot is displayed in the upper half, whereas the bar plot appears in the lower half of the Figure Window. It is recommended that all Figure Windows be closed before proceeding to the next example. Subsequent plots would replace the graph in the lower display region only, or in other words, the last generated graph in a Figure Window. Alternatively, the command

```
clf
```

clears the current figure. This command can be used in larger MATLAB scripts after using the function subplot for multiple plots in a Figure Window.

An important modification to graphs is the scaling of the axis. By default, MATLAB uses axis limits close to the minima and maxima of the data. Using the command axis, however, allows the scale settings to be changed. The syntax for this command is simply axis([xmin xmax ymin ymax]). The command

```
plot(x,y1,'r--')
axis([0 pi -1 1])
```

sets the limits of the x-axis to 0 and π, whereas the limits of the y-axis are set to the default values –1 and +1. Important options of axis are

```
plot(x,y1,'r--')
axis square
```

which makes the x-axis and y-axis the same length and

```
plot(x,y1,'r--')
axis equal
```

which makes the individual tick mark increments on the x-axis and y-axis the same length. The function grid adds a grid to the current plot, whereas the functions title, xlabel and ylabel allow a title to be defined and labels to be applied to the x- and y-axes.

```
plot(x,y1,'r--')
title('My first plot')
xlabel('x-axis')
ylabel('y-axis')
grid
```

4 | MATLAB AS A VISUALIZATION TOOL

These are a few examples how MATLAB functions can be used to edit the plot in the Command Window. More graphics functions will be introduced in the following chapters of this book.

4.8 Generating M-Files to Regenerate Graphs

MATLAB supports various ways of editing all objects in a graph interactively using a computer mouse. First, the *Edit Plot* mode of the Figure Window has to be activated by clicking on the arrow icon or by selecting *Edit Plot* from the *Tools* menu. The Figure Window also contains some other options, such as *Rotate 3D*, *Zoom* or *Insert Legend*. The various objects in a graph, however, are selected by double-clicking on the specific component, which opens the *Property Editor*. The Property Editor allows changes to be made to many properties of the graph such as axes, lines, patches and text objects.

The *Generate M-Files* option enables us to automatically generate the MATLAB code of a figure to recreate a graph with different data. We use a simple plot to illustrate the use of the Property Editor and the Generate M-Files option to recreate a graph.

```
clear

x = 0 : pi/10 : 2*pi;
y1 = sin(x);
plot(x,y1)
```

The default layout of the graph is that of *Figure 4.4*. Clicking on the arrow icon in the *Figure Toolbar* enables the Edit Plot mode. The selection handles of the graph appear, identifying the objects that are activated. Double-clicking an object in a graph opens the *Property Editor*.

As an example, we can use the Property Editor to change various properties of the graph. Double-clicking the gray background of the Figure Window gives access to properties such as *Figure Name*, the *Colormap* used in the figure and the *Figure Color*. We can change this color to light blue, represented by the light blue square in the 4th row and 3rd column of the color chart. Moving the mouse over this square displays the RGB color code [0.7 0.78 1]. Activating the blue line in the graph allows us to change the line thickness to 2.0 and select a 6-point square marker. We can close the Property Editor by clicking on the X in the upper right corner of the Property Editor panel below the graph. Finally, we can deactivate the Edit Plot mode of the Figure Window by clicking on the arrow icon in the Figure Toolbar.

After having made all necessary changes to the graph, the correspond-

ing commands can then be exported by selecting *Generate M-File* from
the *File* menu of the Figure Window. The generated code displays in the
MATLAB Editor.

```
function createfigure(X1, Y1)
%CREATEFIGURE(X1,Y1)
%   X1:   vector of x data
%   Y1:   vector of y data

%   Auto-generated by MATLAB on 06-Oct-2009 17:37:20

% Create figure
figure1 = figure('XVisual',...
    '0x24 (TrueColor, depth 24,...
    RGB mask 0xff0000 0xff00 0x00ff)',...
    'Color',[0.6784 0.9216 1]);

% Create axes
axes('Parent',figure1);
box('on');
hold('all');

% Create plot
plot(X1,Y1,'Marker','square','LineWidth',2);
```

We can then rename the function *createfigure* to *mygraph* and save the file
as *mygraph.m* using *Save As* from the *Editor File* menu.

```
function mygraph(X1, Y1)
%MYGRAPH(X1,Y1)
%   X1:   vector of x data
%   Y1:   vector of y data
(cont'd)
```

The automatically-generated graphics function illustrates how graphics are
organized in MATLAB. The function `figure` first opens a Figure Window.
Using `axes` then establishes a coordinate system, and using `plot` draws
the actual line object. The Figure section in the function reminds us that
the light-blue background color of the Figure Window is represented by the
RGB color coding `[0.702 0.7804 1]`. The Plot section reveals the square
marker symbol used and the line width of 2 points.

The newly-created function `mygraph` can now be used to plot a different data set. We use the above example and

```
clear

x = 0 : pi/10 : 2*pi;
y2 = cos(x);
mygraph(x,y2)
```

The figure shows a new plot with the same layout as the previous plot. The

Generate M-File function of MATLAB can therefore be used to create templates for graphs that can be used to generate plots of multiple data sets using the same layout.

Even though MATLAB provides enormous editing facilities and the Generate M-File function even allows the generation of complex templates for graphs, a more practical way to modify a graph for presentations or publications is to export the figure and import it into a different software such as CorelDraw or Adobe Illustrator. MATLAB graphs are exported by selecting the command *Save as* from the *File* menu or by using the command `print`. This function exports the graph, either as a raster image (e.g., JPEG or GIF) or as a vector file (e.g., EPS or PDF), into the working directory. In practice, the user should check the various combinations of export file formats and the graphics software used for final editing of the graphs.

4.9   Publishing M-Files

A relatively new feature of the software is the option to publish reports on MATLAB projects in various file formats such as HTML, XML, LaTeX and many others. This feature enables you to share your results with colleagues who may or may not have the MATLAB software. The published code includes formatted commentary on the code, the actual MATLAB code and all results of running the code, including the output to the Command Window and all graphs created or modified by the code.

To illustrate the use of the publishing feature, we create a simple example of a commented MATLAB code to compute the sine and cosine of a time vector and display the results as two separate figures. Before we start developing the MATLAB code, we activate *Enable Cell Mode* in the *Cell* menu of the Editor. Whereas single percent signs % are known (from Section 4.6) to initiate comments in MATLAB, we now use double percent signs %% that indicate the start of new cells in the Editor. The *Cell Mode* is a feature in MATLAB that enables you to evaluate blocks of commands by using the buttons *Evaluate Cell*, *Evaluate Cell and Advance* and *Evaluate Entire File* on the *Editor Cell Mode* toolbar. The *Save and Publish* button, which was situated next to the Cell Mode buttons in earlier versions of MATLAB, is now included in the Editor Toolbar emphasizing the importance and popularity of this feature.

We start the Editor by typing `edit` in the Command Window, which opens a new window named *untitled*. An M-file to be published starts with a document title at the top of the file followed by some comments that describe the contents and the version of the script. The subsequent contents of the file include cells of MATLAB code and comments separated by the double percent signs %%.

```
%% Example for Publishing M-Files
% This M-file illustrates the use of the publishing
% feature of MATLAB.
% By Martin Trauth, Feb 8, 2009.

%% Sine Wave
% We define a time vector t and compute the sine y1 of t.
% The results are displayed as linear 2D graph y1 against x.
x = 0 : pi/10 : 2*pi;
y1 = sin(x);
plot(x,y1)
title('My first plot')
xlabel('x-axis')
ylabel('y-axis')

%% Cosine Wave
% Now we compute the cosine y2 of the same time vector and
% display the results.
y2 = sin(x);
plot(x,y2)
title('My first plot')
xlabel('x-axis')
ylabel('y-axis')

%%
% The last comment is separated by the double percent sign
% without text. This creates a comment in a separate cell
% without a subheader.
```

We save the M-file as *myproject.m* and click the *Publish myproject.m* button in the Editor Toolbar. The entire script is now evaluated and the Figure Windows pop up while the script is running. Finally, a window opens up that shows the contents of the published M-file. The document title and subheaders are shown in a dark red font, whereas the comments are in black fonts. The file includes a list of contents with jump links to proceed to the chapters of the file. The MATLAB commands are displayed on gray backgrounds, but the graphs are embedded in the file without the gray default background of Figure Windows. The resulting HTML file can be easily included on a course or project webpage. Alternatively, the HTML file and included graphs can be saved as a PDF-file and shared with students or colleagues.

Recommended Reading

Davis TA, Sigmon K (2005) The MATLAB Primer, Seventh Edition. Chapman & Hall/CRC, London

Etter DM, Kuncicky DC, Moore H (2004) Introduction to MATLAB 7. Prentice Hall, New Jersey

Gilat A (2007) MATLAB: An Introduction with Applications. John Wiley & Sons, New York

Hanselman DC, Littlefield BL (2004) Mastering MATLAB 7. Prentice Hall, New Jersey

Palm WJ (2004) Introduction to MATLAB 7 for Engineers. McGraw-Hill, New York

The MathWorks (2011) MATLAB Getting Started Guide. The MathWorks Inc., Natick, MA

5 Visualizing 2D Data in Earth Sciences

5.1 Introduction

In this chapter we demonstrate advanced two-dimensional visualization techniques in the form of graphical displays of the types of data typically encountered in earth sciences, using MATLAB. The first example displays graphically a temperature and snow accumulation time series for the last 20,000 years from the GISP2 ice core data presented by R.B. Alley (2000), in a single plot with an x-axis and two y-axes (Section 5.2). Section 5.3 introduces the use of bar plots for displaying histograms, in which the temperature data used in the previous example is divided over equally spaced temperature intervals (called bins, or classes) and the counts per bin (the number of data points that fall within each bin) are displayed as a bar plot. The same frequency distribution for temperature values is then displayed as two- and three-dimensional pie charts, and the use of transparency is introduced (Section 5.4). The visualization of directional data, which is particularly important in earth sciences, is illustrated in Section 5.5 although MATLAB is certainly not the best software for displaying data of this type. The last two sections of the chapter then introduce two specialized MATLAB scripts for displaying multiproxy data, such as pollen or microfossil records, and for visualizing stratigraphic logs of sedimentary sequences.

5.2 Line Graphs: Plotting Time Series in Earth Sciences

Our first example demonstrates how to create line graphs with MATLAB. We first load the GISP2 data for temperature and snow accumulation variations (from R.B. Alley 2000) from the two files created in Chapter 3.

```
clear
temp = load('icecore_temperature_data.txt');
accum = load('icecore_snowaccumulation_data.txt');
```

We then generate a simple line graph of the temperature and snow accumulation data, as demonstrated in Chapter 4.

M. H. Trauth and E. Sillmann, *MATLAB® and Design Recipes for Earth Sciences*, DOI: 10.1007/978-3-642-32544-1_5, © Springer-Verlag Berlin Heidelberg 2013

```
plot(temp(:,1),temp(:,2))
```

This command opens a *Figure Window* named *Figure 1*, with a gray background, an *x*-axis ranging from 0 to 50 kyr, a *y*-axis ranging from –55 to –25 deg C, and a blue line. Using the command axis allows the scale settings to be changed. The syntax for this command is simply axis([xmin xmax ymin ymax]). The command axis([0 20 -60 -25]) sets the limits of the *x*-axis to 0 and 20 kyr and those of the *y*-axis to –60 and –25 deg C. The functions title, xlabel and ylabel allow a title to be defined and labels to be applied to the *x*- and *y*-axes.

```
plot(temp(:,1),temp(:,2))
axis([0 20 -60 -25])
title('GISP Ice Core Temperature Data')
xlabel('Age- Thousands of Years Before the Present')
ylabel('Temperature- Degrees Celsius')
```

Advanced editing of graphics requires an understanding of the MATLAB requirement for graphics to be organized as hierarchical suites of graphics objects. Put simply, the hierarchies of graphical objects include three main layers: *figure window*, *axes*, and *plot objects*. The objects within each layer have a fixed set of properties, most of which can be modified to alter the appearance of a graph. The values of each set of properties can be queried and most of these values can be modified. The values for the properties of the current figure window can be queried by

```
get(gcf)
```

Taking one example out of the long list of properties, the value of the property Color, which represents the light gray color of the figure window, is [0.8 0.8 0.8]. Typing

```
get(gca)
```

gives us the properties of the current axes. As can be seen, the limits of the *x*-axis are given by XLim = [0 20] and those of the *y*-axis are given by YLim = [-60 -25]. All graphics functions ascribe a *handle* to each graphics object, which we can store as variables for each type of object. Examples of such variables are figure1, axes1, and line1, which can be used to access the graphics properties for the figure window, the axes and the line in a graph.

```
figure1 = figure('Color',[1 1 1]);
axes1 = axes('XLim',[0 20],...
    'XDir','reverse',...
```

```
        'YLim',[-60 -20],...
        'Box','on');
line1 = line(temp(:,1),temp(:,2),...
        'Color',[0.1 0.3 0.8]);
title1 = title('GISP Ice Core Temperature Data');
xlabel1 = xlabel('Age- Thousands of Years Before the Present');
ylabel1 = ylabel('Temperature- Degrees Celsius');
```

This series of commands sets the background color of the figure window to red, green, blue (RGB) values of [1 1 1], which is white. The limits of the *x*- and *y*-axes are then defined, with the *x*-axis being reversed and the two open sides of the graph closed into form a box. The line color is next defined as [0.1 0.3 0.8], which is a medium blue. We can access the values of an individual graphics property, for instance the Color property of the line object line1, with

```
get(line1,'Color')
```

resulting in the output

```
ans =
    0.1000    0.3000    0.8000
```

which respresents the RGB values [0.1 0.3 0.8] of the line in the graph. We can change the color by typing, for instance

```
set(line1,'Color',[0.8 0.3 0.1])
```

Similarly, we can modify the title of the graph by typing

```
set(title1,...
    'String',...
    ['GISP Ice Core Temperature Data, Alley, R.B. 2000']);
```

We can also use the command set to modify any graphics property, as follows:

```
figure1 = figure('Color',[1 1 1]);
axes1 = axes('XLim',[0 20],...
    'XDir','reverse',...
    'YLim',[-60 -20],...
    'Box','on');
line1 = line(temp(:,1),temp(:,2),...
    'Color',[0.1 0.3 0.8]);
set(get(axes1,'Title'),...
    'String','GISP Ice Core Temperature Data')
set(get(axes1,'XLabel'),...
    'String','Age- Thousands of Years Before the Present')
set(get(axes1,'YLabel'),...
    'String','Temperature- Degrees Celsius')
```

This script creates a similar plot to the previous one but uses the object handle `axes1` to modify the title, x-label and y-label.

We now wish to plot both the temperature and the snow accumulation data on a single plot, each represented by different colored lines, using the function `plotyy`. This function creates plots with different y-axes on the left and the right of the graph, each with different tick labels, and with the axis labels and colors corresponding to the colors of the lines.

```
plotyy(temp(:,1),temp(:,2),accum(:,1),accum(:,2));
```

The figure window properties can be accessed as they were previously, but any modification of the axis and line properties requires the introduction of separate object handles for each of the two lines. The object handle `axes1` now has two values, one for each of the two y-axes, whereas two separate object handles are used for the two plotted lines, these being `line1` and `line2`. We can use a script similar to that used in the previous example to modify the graph:

```
figure1 = figure('Color',[1 1 1]);
[axes1,line1,line2] = ...
    plotyy(temp(:,1),temp(:,2),accum(:,1),accum(:,2));
set(get(axes1(1),'Title'),...
    'String','GISP Ice Core Temperature and Accumulation Data')
set(get(axes1(1),'XLabel'),...
    'String','Age- Thousands of Years Before the Present')
set(get(axes1(1),'YLabel'),...
    'String','Temperature- Degrees Celsius')
set(get(axes1(2),'YLabel'),...
    'String','Snow Accumulation- meters/year')
set(axes1(1),'YColor',[0.8 0.3 0.1],...
    'XLim',[0 20],...
    'XDir','reverse',...
    'YLim',[-60 -20]);
set(axes1(2),'YColor',[0.1 0.3 0.8],...
    'XLim',[0 20],...
    'XDir','reverse',...
    'YLim',[0 0.4]);
set(line1,'Color',[0.8 0.3 0.1])
set(line2,'Color',[0.1 0.3 0.8])
```

In this book we try to display all graphs in a similar layout. Our graphics template has the following settings: the width of the x-axis is 10 cm, the height of the y-axis is 8 cm, the line thickness of the axes is 0.6 pt (0.6 points), the line thickness of curves is 0.5 pt, the axis labels and title are in a sans serif font such as *Helvetica*, the font size of the axis labels is 8 pt, and the font size of the title is 10 pt (Fig. 5.1). The corresponding MATLAB script reads as follows:

Fig. 5.1 Line plots of temperature and snow accumulation variations from the GISP2 data (from R.B. Alley 2000) derived from the two files created in Chapter 3.

```
figure1 = figure('Color',[1 1 1]);
[axes1,line1,line2] = ...
    plotyy(temp(:,1),temp(:,2),accum(:,1),accum(:,2));
set(get(axes1(1),'Title'),...
    'String','GISP Ice Core Temperature and Accumulation Data',...
    'FontSize',10);
set(get(axes1(1),'XLabel'),...
    'String','Age- Thousands of Years Before the Present',...
    'FontSize',8);
set(get(axes1(1),'YLabel'),...
    'String','Temperature- Degrees Celsius',...
    'FontSize',8);
set(get(axes1(2),'YLabel'),...
    'String','Snow Accumulation- meters/year',...
    'FontSize',8);
set(axes1(1),...
    'YColor',[0.8 0.3 0.1],...
    'XLim',[0 20],...
```

```
        'XDir','reverse',...
        'YLim',[-60 -20],...
        'Units','Centimeters',...
        'Position',[3 2 10 8],...
        'LineWidth',0.6,...
        'FontSize',8);
    set(axes1(2),'YColor',[0.1 0.3 0.8],...
        'XLim',[0 20],...
        'XDir','reverse',...
        'YLim',[0 0.4],...
        'Units','Centimeters',...
        'Position',[3 2 10 8],...
        'LineWidth',0.6,...
        'FontSize',8);
    set(line1,'Color',[0.8 0.3 0.1],...
        'LineWidth',0.5)
    set(line2,'Color',[0.1 0.3 0.8],...
        'LineWidth',0.5)
```

This script can be easily modified to suit to the reader's own requirements. Other graphics properties can also be included to change the final layout of the line graph. Having completed our line graph, we save the result in a file for further editing in a vector graphics program (see Chapter 8).

```
print -depsc2 icecore_lineplot_vs1_matlab.eps
```

The figure window needs to be active when saving an image to a file using print. The preferred format to use for saving vector graphics such as the one in our example is the *Encapsulated Level 2 Color PostScript* (EPSC2) format. Note that the axes of the plot consist of several superimposed lines that cause color interferences. A possible way around this problem is described in Chapter 8.

5.3   Bar Graphs: Plotting Histograms in Earth Sciences

Our second example demonstrates how to create bar graphs with MATLAB. Again, we use the temperature data from R.B. Alley (2000) contained in the file *icecore_temperature_data.txt*.

```
clear
temp = load('icecore_temperature_data.txt');
```

In statistical analyses, bar graphs (or plots), or histograms, are used to visualize frequency distributions of univariate data sets that have been organized into separate intervals, bins, or classes. We use a histogram to display the frequency distribution of the temperature values. By default, the function hist divides the range of the data into ten bins, counts the number of observations within each bin, and displays the frequency distribution as a bar graph.

```
v = -55 : 2.5 : -25;
hist(temp(:,2),v)
```

These two lines of code create a histogram with bin centers at 2.5 deg C intervals from -55 to -25 deg C. The midpoints of the default bin v and the number of observations per bin n can be accessed using

```
[n,v] = hist(temp(:,2),v);
```

After calculating n and v, hist uses bar to draw a bar plot. The default bar width of the function bar is 0.8 (or 80 %), while hist uses a bar width of 1.0 (or 100 %).

```
bar(v,n)
```

We can modify the bar width by including the width as a graphics property and inserting its value within the function input

```
bar(v,n,'BarWidth',1.0)
```

Alternatively, we can change the value of the property by accessing the graphics object handle, as described in the previous section. We can first get the value of the BarWidth property by typing

```
h = bar(v,n);
get(h,'BarWidth')
```

The value of the BarWidth property can then be changed by typing

```
h = bar(v,n);
set(h,'BarWidth',1.0)
```

We again display the graphs in a similar layout to that described previously (Fig. 5.2). The corresponding MATLAB script reads as follows:

```
figure1 = figure('Color',[1 1 1]);
axes1 = axes('Units','Centimeters',...
    'Position',[3 2 10 8],...
    'Box','on',...
    'LineWidth',0.6,...
    'FontSize',8);
bar1 = bar(v,n,...
    'BarWidth',1.0,...
    'LineWidth',0.5,...
    'FaceColor',[0.8 0.3 0.1]);
title1 = title('GISP Ice Core Temperature Data',...
```

```
    'FontSize',10);
xlabel1 = xlabel('Temperature- Degrees Celsius',...
    'FontSize',8);
ylabel1 = ylabel('Frequency',...
    'FontSize',8);
```

Once again this script can be easily modified. Having completed our bar plot, we save the result in a file for further editing in vector graphics software (see Chapter 8).

```
print -depsc2 icecore_bargraph_vs1_matlab.eps
```

As stated previously, the figure window needs to be active while saving the image to a file, and the preferred format to use for saving vector graphics such as the one in our example is the EPSC2 format.

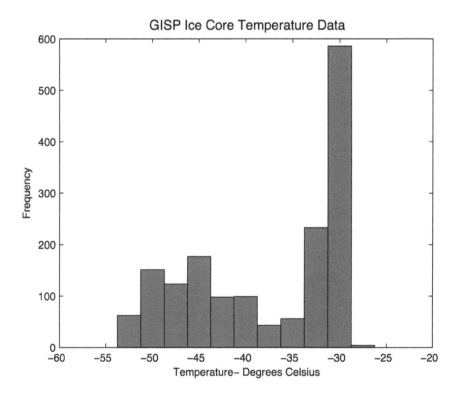

Fig. 5.2 Bar plot of the frequency distribution of temperature values from the GISP2 data (from R.B. Alley 2000) derived from the file created in Chapter 3.

5.4   Pie Charts: Illustrating Proportion in Earth Sciences

Our third example demonstrates how to create pie charts with MATLAB. Again, we use the temperature data from R.B. Alley (2000) contained in the file *icecore_temperature_data.txt*.

```
clear
temp = load('icecore_temperature_data.txt');
```

We then again calculate the frequency distribution of the temperature data:

```
v = -55 : 2.5 : -25;
[n,v] = hist(temp(:,2),v);
```

We can now create a pie chart of the frequency distribution. This chart does not provide any information on the actual values that contribute to each slice of the pie. The variable n contains two empty bins, one at either end, which are ignored when plotting and cause the message *Warning: Ignoring non-positive data in pie chart.*

```
pie(n)
```

The function pie basically scales the data to 100 % and plots it as slices of a pie. The total number of temperature values in our example is sum(n)=1632. All values in n are therefore divided by 1632 and multiplied by 100 to obtain percentages. As you can see, the largest value is 35.9069, which is rounded to the nearest integer, which is 36, for the plot label.

```
percentages = 100*n/1632
```

The function pie also allows some slices to be offset from the center, using the vector explode to identify the slices to be offset by marking them with a 1, while the slices that are not offset are marked with a 0.

```
explode = zeros(size(n));
explode(1,5:7) = [1 1 1];
pie(n,explode)
```

We can also plot the frequency distribution in a 3D pie chart. This will be the first 3D plot that we create in this book. We will learn more about 3D charts in the next chapter of this book.

```
pie3(n)
```

We again display the graphs in our standard layout. The corresponding MATLAB script reads as follows:

```
figure1 = figure('Color',[1 1 1]);
axes1 = axes('Units','Centimeters',...
    'Position',[3 2 10 8],...
    'LineWidth',0.6,...
    'FontSize',8);
pie1 = pie3(n);
```

As before, this script can be easily modified. Having completed our bar plot, we save the result in a file for further editing in vector graphics software.

```
print -depsc2 icecore_piechart_vs1_matlab.eps
```

Again, the figure window needs to be active while saving the image to a file and the preferred format to use for saving vector graphics such as the one in our example is the EPSC2 format.

A useful feature of MATLAB graphics is that of transparency. We use the function `alpha` to make the pie chart transparent, using a value of 0.3 (Fig. 5.3), where 1 is opaque and 0 is clear.

```
figure1 = figure('Color',[1 1 1]);
axes1 = axes('Units','Centimeters',...
    'Position',[3 2 10 8],...
    'LineWidth',0.6,...
    'FontSize',8);
pie1 = pie3(n);
alpha(0.3)
```

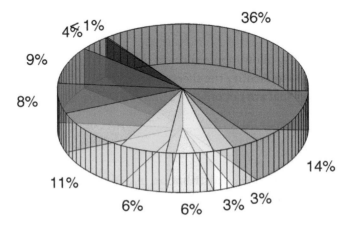

Fig. 5.3 Pie chart of the temperature data from R.B. Alley (2000) contained in the file *icecore_temperature_data.txt*. We use the function `alpha` to make the pie chart transparent with a value of 0.3, where 1 is opaque and 0 is clear.

Unfortunately, graphs using transparency are not exported as vector graphics. Trying to export the pie chart as an EPS file creates a raster or pixel version of the graph.

```
print -depsc2 icecore_piechart_vs2_matlab.eps
```

We therefore export the transparent pie chart as a JPEG file

```
print -djpeg70 -r600 icecore_piechart_vs3_matlab.jpg
```

with a quality level of 70 % and a resolution of 600 dpi. Alternatively, we can export the non-transparent version of the pie chart and introduce transparency using vector graphics software (Chapter 8).

5.5 Rose Diagrams: Plotting Directional Data

The classic way to display directional data is the rose diagram. A rose diagram is a histogram for angle measurements. In contrast to a bar histogram, in which the height of the bars is proportional to the frequency, a rose diagram comprises sectors of a circle in which the radius of each sector is proportional to the frequency. We use synthetic data to illustrate two types of rose diagrams for displaying directional data. We first load a set of directional data from the file *directional_data.txt*.

```
clear

data_degrees_1 = load('directional_data.txt');
```

This data set contains forty measurements of angles, in degrees. We use the function `rose(az,nb)` to display the data. This function plots an angle histogram for the angles `az` in radians, where `nb` is the number of classes. However, since the original data are in degrees, we need to convert all measurements to radians before we plot the data.

```
data_radians_1 = pi*data_degrees_1/180;
rose(data_radians_1,12)
```

The function `rose` counts in a counterclockwise direction, with zero degrees lying along the *x*-axis of the coordinate graph. In geosciences, however, the opposite applies and degrees are counted in a clockwise direction from due north, which is zero degrees. The command `view` rotates the plot by +90° (the azimuth) and mirrors the plot by –90° (the elevation).

```
rose(data_radians_1,12)
view(90,-90)
```

VISUALIZING 2D DATA IN EARTH SCIENCES

5

The area of the arc segments increases with class frequency. In a final modification the rose diagram is therefore scaled to the square root of the class frequency. The function `rose` does not allow plotting of the square root of the frequencies by default, but the corresponding file `rose.m` can be easily modified as follows. After the histogram of the angles has been computed (in line 58) by using the function `histc`, add a line with the command nn = sqrt (nn) ;, which computes the square root of the frequencies nn. Save the modified function as file `rose_sqrt.m` and apply the new function to the data set.

```
rose_sqrt(data_radians_1,12)
view(90,-90)
```

This plot now satisfies all the geoscience conventions. We again display the graph in our standard layout (Fig. 5.4). The corresponding MATLAB script reads as follows:

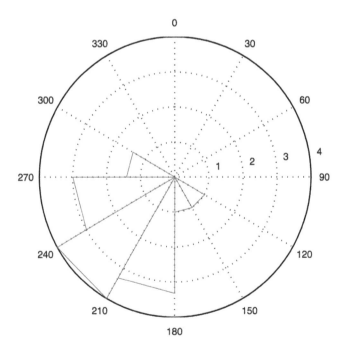

Fig. 5.4 Rose diagram of directional data from the file *directional_data.txt*.

```
figure1 = figure('Color',[1 1 1]);
axes1 = axes('Units','Centimeters',...
     'Position',[3 2 10 8],...
     'LineWidth',0.6,...
     'FontSize',8);
rose1 = rose_sqrt(data_radians,12);
set(rose1,'Color',[0.8 0.3 0.1])
view(90,-90)
```

As before, this script can be easily modified. Having completed our rose diagram, we save the result in a file for further editing in vector graphics software (see Chapter 8).

```
print -depsc2 icecore_rosediagram_vs1_matlab.eps
```

Once again, the figure window needs to be active while saving the image to a file and the preferred format to use for saving vector graphics such as the one in our example is the EPSC2 format.

5.6 Multiplots: Plotting Scaled Multiple Area Graphs

This specialized example creates several plots in a single figure window. It plots patch, line, or bar plots of several pollen or microfossil records, all on axes with the same scales. We first need to create some synthetic data. The first column contains the age vector, while all other data are uniformly-distributed pseudo random numbers. The total of all values within each column is 100. We can change the number of data points m.

```
clear
m = 160;
data(:,1)  = (1 : m)';
data(:,2:15) = rand(m,14);
data(:,2)  = 5*data(:,2);
data(:,3)  = 5*data(:,3);
data(:,4)  = 20*data(:,4);
data(:,5)  = 5*data(:,5);
data(:,6)  = 5*data(:,6);
data(:,7)  = 3*data(:,7);
data(:,8)  = 5*data(:,8);
data(:,9)  = 5*data(:,9);
data(:,10) = 20*data(:,10);
data(:,11) = 15*data(:,11);
data(:,12) = 5*data(:,12);
data(:,13) = 3*data(:,13);
data(:,14) = 2*data(:,14);
data(:,15) = 2*data(:,15);
```

The synthetic data is then saved to the binary file *multipledata_data.mat*.

```
save multipledata_data.mat data
```

We now clear the workspace and import the data from the binary file *mul-tipledata_data.mat*.

```
clear
load multipledata_data.mat
```

Alternatively, we can import any other data set, with an arbitrary number of data points and variables, provided the total comes to 100 %. The multiplot first creates a figure window with the same size as the screen. It then plots all records at the same scale (Fig. 5.5).

```
scrsz = get(0,'ScreenSize');
scrsz = 0.7*scrsz;
colorcode = [0.8 0.3 0.1];
sf = 0.75;
[m,n] = size(data);
data(2:m+1,:) = data;
data(1,:) = zeros(1,n);
data(1,1) = data(2,1);
data(m+2,:) = zeros(1,n);
data(m+2,1) = data(m,1);
figure1 = figure(...
    'Position',[1 scrsz(4) scrsz(3) scrsz(4)],...
    'Color',[1 1 1]);
for i = 1:n-1
    if i == 1
    xcoord = max(data(:,2))/100;
    axes1 = axes(...
        'Parent',figure1,...
        'YMinorTick','on',...
        'XMinorTick','on',...
        'TickDir','out',...
        'Position',[0.1 0.1 sf*xcoord 0.6],...
        'FontSize',8,...
        'XLim',[0 max(data(:,2))],...
        'YLim',[min(data(:,1)) max(data(:,1))]);
    box(axes1,'off');
    hold(axes1,'all');
    patch(data(:,2),data(:,1),colorcode)
    % plot(data(:,2),data(:,1))
    % barh(data(2:end-1,1),data(2:end-1,2))
    ylabel('Year')
    xlabel('%')
    xcoord = xcoord + 0.01;
    else
    axes1 = axes(...
        'Parent',figure1,...
        'YTick',zeros(1,0),...
        'FontSize',8,...
        'YMinorTick','on',...
        'XMinorTick','on',...
```

Fig. 5.5 Multiple patch plots displaying pollen and microfossil records, all plotted at the same scale.

```
            'TickDir','out',...
            'Position',[sf*xcoord+0.1 0.1 ...
                sf*max(data(:,i+1))/100 0.6],...
            'XLim',[0 max(data(:,i+1))],...
            'YLim',[min(data(:,1)) max(data(:,1))]);
        box(axes1,'off');
        hold(axes1,'all');
        patch(data(:,i+1),data(:,1),colorcode)
        % plot(data(2:end-1,i+1),data(2:end-1,1))
        % barh(data(2:end-1,1),data(2:end-1,i+1))
        xlabel('%')
        xcoord = xcoord + max(data(:,i+1))/100 + 0.01;
        end
    end
```

We can change the multiplier for scrsz between 0 and 1 to match the figure size with the screen size. We can also change colorcode to modify the color of the patches. The variable sf scales the graph with respect to the figure window. Instead of a patch plot, the graphics function plot for a line plot and barh for a horizontal bar plot are also available. To use either of these we need to uncomment the patch, the plot or the barh command by removing the % signs at the beginning of the corresponding lines.

As in all previous examples, this script can be easily modified. Having completed our bar plot, we save the result in a file for further editing in vector graphics software (see Chapter 8).

```
print -depsc2 multipledata_multiplot_vs1_matlab.eps
```

As previously stated, the figure window needs to be active while saving the image to a file, and the preferred format to use for saving vector graphics such as the one in our example is the EPSC2 format.

5.7   Stratplots: Plotting Stratigraphic Columns

The second specialized example introduces a MATLAB script to draw a stratigraphic column with either angular or rounded corners, labels, and several data curves. We first load the data from a binary file. The file contains the data matrix and a variable containing labels such as rock types.

```
clear
load stratigraphiccolumn_data.mat
```

We need to flip the data array data and the label array litho in up/down direction if the first line in the data array represents the lowermost sediment layer, as this needs to be displayed at the bottom of the graph.

```
data = flipud(data);
litho = flipud(litho);
```

Next, we add a zero to either end in order to close the patches in the plot.

```
[n,m] = size(data);
datanew(1,:) = zeros(1,m);
for i = 1:n
    datanew(i+1,:) = data(i,:);
end
data = datanew;
data(n+2,:) = zeros(1,m);
```

We define specific variables, such as t for the layer thickness, w for the degree of weathering, l for the type of lithology, and ct for the cumulative thickness, to plot the data.

```
t = data(:,1);
w = data(:,2);
wc = data(:,2)./max(data(:,2));
l = data(:,3);
lc = data(:,3)./max(data(:,3));
ct = cumsum(t);
```

Finally, we display the stratigraphic column as a sequence of rectangular colored layers in which the color represents the sediment type. The script first creates a white figure window, then creates a stratigraphic column with angular corners using fill, labels the layers, and creates two curves using the function patch.

```
figure1 = figure('Color',[1 1 1]);

% PART 1 Stratigraphic column

axes('position',[0.15 0.1 0.1 0.8],...
    'YLim',[0 max(ct)],...
    'Color','None',...
    'TickDir','out',...
    'XTickLabel',[],...
    'XTick',[])
hold on
for i=2:length(ct)
    layercolor = [0.8 lc(i) 0.1];
    X=[0 0 w(i) w(i)];
    Y=[ct(i-1) ct(i) ct(i) ct(i-1)];
    fill((X).^.5,Y,layercolor)
end
hold off

% PART 2: Labels

axes('position',[0.25 0.1 0.1 0.8],...
```

```
    'YLim',[0 max(ct)],...
    'Color','None',...
    'Visible','off')
for i=1:length(ct)-2
    text(0.35,ct(i+1)-t(i+1)./2,litho(i,:),...
        'Fontsize',6)
end

% PART 3: Patch 1

axes('position',[0.4 0.1 0.1 0.8],...
    'YLim',[0 max(ct)],...
    'Color','None',...
    'TickDir','out',...
    'XTickLabel',[],...
    'XTick',[])
patch(data(:,4),ct,[0.8 0.3 0.1])

% PART 4: Patch 2

axes('position',[0.55 0.1 0.1 0.8],...
    'YLim',[0 max(ct)],...
    'Color','None',...
    'TickDir','out',...
    'XTickLabel',[],'XTick',[])
patch(data(:,5),ct,[0.8 0.3 0.1])
```

Having completed our plot, we save the result in a file for further editing in vector graphics software (see Chapter 8).

```
print -depsc2 stratigraphiccolumn_stratplot_vs1_matlab.eps
```

Once again, the figure window needs to be active while saving the image to a file and the preferred format to use for saving the vector graphics such as the one in our example is the EPSC2 format.

A modification of the previous graph is a stratigraphic column with rounded corners to the layers. The script first creates a white figure window, then creates a stratigraphic column with rounded corners using `fill`, labels the layers, and creates two curves using the function `patch`. The script is slightly more complicated than that for the angular version since the rounded corners differ depending on the width shown for the overlying and underlying layers. The rounding is either convex or concave depending on the neighboring layer, or not rounded if the neighboring layers has the same width. The script for labeling the layers and for creating the two curves using `patch` is very similar to those described in the previous sections (Fig. 5.6).

```
figure1 = figure('Color',[1 1 1]);
```

```
% PART 1 Stratigraphic column

axes('position',[0.15 0.1 0.1 0.8],...
    'YLim',[0 max(ct)],...
    'Color','None',...
    'TickDir','out',...
    'XTickLabel',[],...
    'XTick',[])
hold on
for i=2:length(ct)-1
  layercolor = [0.8 lc(i) 0.1];

% If loop, w=1 narrow, w=2 wide, w=0 bottom or top
if   w(i)==2&w(i-1)==1&w(i+1)==1
  X=[0 0 w(i)-0.2 w(i) w(i) w(i)-0.2];
  Y=[ct(i-1) ct(i) ct(i) ct(i)-0.2*mean(t) ...
      ct(i-1)+0.2*mean(t) ct(i-1)];
  fill((X).^.5,Y,layercolor)
elseif w(i)==1 & w(i-1)==2 & w(i+1)==2
  X=[0 0 w(i)+0.2 w(i) w(i) w(i)+0.2];
  Y=[ct(i-1) ct(i) ct(i) ct(i)-0.2*mean(t) ...
      ct(i-1)+0.2*mean(t) ct(i-1)];
  fill((X).^.5,Y,layercolor)
elseif w(i)==1&(w(i-1)==1&w(i+1)==1)|(w(i-1)==0&w(i+1)==1)
  X=[0 0 w(i) w(i)];
  Y=[ct(i-1) ct(i) ct(i) ct(i-1)];
  fill((X).^.5,Y,layercolor)
elseif w(i)==1&w(i-1)==1&w(i+1)==0
  X=[0 0 w(i) w(i)];
  Y=[ct(i-1) ct(i) ct(i) ct(i-1)];
  fill((X).^.5,Y,layercolor)
elseif w(i)==2&(w(i-1)==2&w(i+1)==2)|(w(i-1)==0&w(i+1)==2)
  X=[0 0 w(i) w(i)];
  Y=[ct(i-1) ct(i) ct(i) ct(i-1)];
  fill((X).^.5,Y,layercolor)
elseif w(i)==2&w(i-1)==2&w(i+1)==0
  X=[0 0 w(i) w(i)];
  Y=[ct(i-1) ct(i) ct(i) ct(i-1)];
  fill((X).^.5,Y,layercolor)
elseif w(i)==2&(w(i-1)==2&w(i+1)==1)|(w(i-1)==0&w(i+1)==1)
  X=[0 0 w(i)-0.2 w(i) w(i)];
  Y=[ct(i-1) ct(i) ct(i) ct(i)-0.2*mean(t) ct(i-1)];
  fill((X).^.5,Y,layercolor)
elseif w(i)==2&(w(i-1)==1&w(i+1)==2)|(w(i-1)==1&w(i+1)==0)
  X=[0 0 w(i) w(i) w(i)-0.2];
  Y=[ct(i-1) ct(i) ct(i) ct(i-1)+0.2*mean(t) ct(i-1)];
  fill((X).^.5,Y,layercolor)
elseif w(i)==1&w(i-1)==1&w(i+1)==2
  X=[0 0 w(i)+0.2 w(i) w(i)];
  Y=[ct(i-1) ct(i) ct(i) ct(i)-0.2*mean(t) ct(i-1)];
  fill((X).^.5,Y,layercolor)
elseif w(i)==1&w(i-1)==2&w(i+1)==1
  X=[0 0 w(i) w(i) w(i)+0.2];
  Y=[ct(i-1) ct(i) ct(i) ct(i-1)+0.2*mean(t) ct(i-1)];
  fill((X).^.5,Y,layercolor)
end
```

```
end

% PART 2: Labels

axes('position',[0.25 0.1 0.1 0.8],...
     'YLim',[0 max(ct)],...
     'Color','None',...
     'Visible','off')
for i=1:length(ct)-2
    text(0.35,ct(i+1)-t(i+1)./2,litho(i,:),...
        'Fontsize',6)
end

% PART 3: Patch 1

axes('position',[0.4 0.1 0.1 0.8],...
     'YLim',[0 max(ct)],...
     'Color','None',...
     'TickDir','out',...
     'XTickLabel',[],...
     'XTick',[])
patch(data(:,4),ct,[0.8 0.3 0.1])

% PART 4: Patch 2

axes('position',[0.55 0.1 0.1 0.8],...
     'YLim',[0 max(ct)],...
     'Color','None',...
     'TickDir','out',...
     'XTickLabel',[],'XTick',[])
patch(data(:,5),ct,[0.8 0.3 0.1])
```

Having completed our plot, we save the result in a file for further editing in vector graphics software (see Chapter 8).

```
print -depsc2 stratigraphiccolumn_stratplot_vs2_matlab.eps
```

As always, the figure window needs to be active while printing the image to a file and the preferred format to use for saving vector graphics such as the one in our example is again the EPSC2 format.

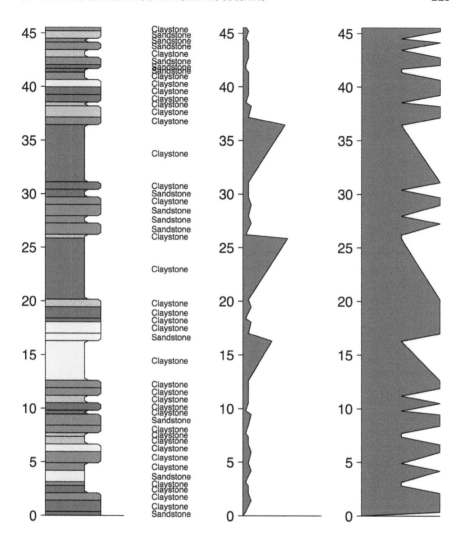

Fig. 5.6 Stratigraphic plot using a MATLAB script to draw a stratigraphic column with rounded corners, labels and several data curves.

6 Visualizing 3D Data in Earth Sciences

6.1 Introduction

Most data in earth sciences are spatially distributed, either as *vector data*, (points, lines, polygons) or as *raster data* (gridded topography). Vector data are generated by digitizing map objects such as drainage networks or out-lines of lithologic units. Raster data can be obtained directly from a satellite sensor output, but gridded data can also, in most cases, be interpolated from irregularly-distributed field samples (gridding).

This chapter demonstrates advanced three-dimensional visualization techniques through graphical displays of the types of data typically encountered in earth sciences, using MATLAB. The following section introduces the use of vector data, by using coastline data as an example (Section 6.2). The acquisition and handling of raster data are then illustrated using digital topographic data (Sections 6.3 to 6.5). The availability and use of digital elevation data have increased considerably since the early 1990s. One of the first publicly available data sets for topography and bathymetry was the ETOPO5 data set, which had a resolution of 5 arc minutes (≈ 9 km). In October 2001 this was replaced by ETOPO2, which has a resolution of 2 arc minutes (≈ 4 km), and more recently (in 2009) by ETOPO1 with a resolution of 1 arc minutes (≈ 2 km). There is also a data set for topography called GTOPO30, which was completed in 1996 and has a horizontal grid spacing of 30 arc seconds (≈ 1 km). Following the Shuttle Radar Topography Mission (SRTM) in 2000, the 30 and 90 m resolution SRTM data replaced the older data sets in many scientific studies. Since 2009, the ASTER Global Digital Elevation Map, and its update released in 2011, has provided an alternative data set with a similar, 30 m resolution. In contrast to airborne data, most land-based earth science data are collected on irregular sampling patterns. Access to rocks is often restricted to natural outcrops such as shoreline cliffs and the walls of a gorge, or anthropogenic outcrops such as road cuttings and quarries. Section 6.6 is on interpolating such unevenly-spaced data and displaying the results in two- and three-dimensional graphs.

This chapter requires the Mapping Toolbox, although most graphics

M. H. Trauth and E. Sillmann, *MATLAB® and Design Recipes for Earth Sciences*, DOI: 10.1007/978-3-642-32544-1_6, © Springer-Verlag Berlin Heidelberg 2013

routines used in our examples can be easily replaced by standard MATLAB functions. An alternative and useful mapping toolbox by Rich Pawlowicz (Earth and Ocean Sciences, at the University of British Columbia) is available from

```
http://eos.ubc.ca/~rich/
```

The handling and processing of large spatial data sets also requires a computing system with at least 2 GB physical memory.

6.2   The GSHHS Shoreline Data Set

The *Global Self-consistent, Hierarchical, High-resolution Shoreline* (GSHHS) data set is an amalgamation of two public domain databases by Paul Wessel (SOEST, University of Hawaii, Honolulu, HI) and Walter Smith (NOAA Laboratory for Satellite Altimetry, Silver Spring, MD). The shoreline vector data can be downloaded as MATLAB vector data from the web page of the US National Geophysical Data Center (NGDC):

```
http://ngdc.noaa.gov/mgg/coast/
```

We first need to define the geographic range of interest in decimal degrees, with west and south denoted by a negative sign. For example, the north-east coast of Africa would be displayed between the latitudes of 0 and +15 degrees and between the longitudes of +35 and +55 degrees. We then need to choose the coastline database from which the data is to be extracted. An example is the *World Data Bank II*, which provides maps at a scale 1:2,000,000. The compression method is then set to *None* for the ASCII data and the coast format option is set to *MATLAB*. The resulting Generic Mapping Tool (GMT) map and a link to the raw text data can be displayed by pressing the *Submit - Extract the Coastline File*. Finally, by opening the 340 KB text file on a browser, the data can be saved into a new file called *coastline_data.txt*. The two columns in this file represent the *longitude/latitude* coordinates of polygons or coastline segments that are separated by NaNs.

```
NaN            NaN
42.892067 0.000000
42.893692 0.001760
NaN            NaN
42.891052 0.001467
42.898093 0.007921
42.904546 0.013201
42.907480 0.016721
42.910414 0.020828
42.913054 0.024642
```

```
42.915987 0.028749
42.918921 0.032562
42.922441 0.035789
(cont'd)
```

The NaNs perform two functions: they provide a means of identifying break points in the data and they serve as pen-up commands when plotting vector maps. The shorelines can be displayed by using:

```
clear

data = load('coastline_data.txt');

plot(data(:,1),data(:,2),'k'), axis equal
xlabel('Longitude'), ylabel('Latitude')
```

More advanced plotting functions are contained in the Mapping Toolbox, which allows alternative versions of this plot to be generated:

```
axesm('MapProjection','lambert', ...
      'MapLatLimit',[0 15], ...
      'MapLonLimit',[35 55], ...
      'Frame','on', ...
      'MeridianLabel','on', ...
      'ParallelLabel','on');
plotm(data(:,2),data(:,1),'k');
```

Note that the input for `plotm` is given in the order *longitude*, followed by the *latitude*, i.e., the second column of the data matrix is entered first. In contrast, the function `plot` requires an *xy* input, i.e., the first column is entered first. The function `axesm` defines the map axis and sets various map properties such as the map projection, the map limits and the axis labels.

The shoreline data set is a polygon with 17,703 vertices. This large number causes problems when processing the polygon using vector graphics software. A maximum of 8,000 vertices, but preferably only 4,000, can be processed. To demonstrate the effect of large numbers of vertices, we create a vector graph of the full data set of 17,703 vertices, and then we *decimate* the data set by a factor of 2 and a factor of 4, for comparison. Within the data set, pairs of NaN represent pen up/pen down breaks of the polygon. Simply decimating the the data set by `data = data(1:2:end,:)` would eliminate some of the NaNs and might therefore link polygons representing islands and lakes. A simple script overcomes this problem by creating a second data set `newdata` in which the data is shifted by one data point with respect to the original data set. We then have two data sets with neighboring NaNs. Replacing the real data point in `data` by a NaN if there is one in `newdata` produces neighboring pairs of NaNs in the original data set `data`. We

6 | VISUALIZING 3D DATA IN EARTH SCIENCES

can then decimate the data set by using data(1:2:end, :). To decimate the data set by a factor of 4, we run the script a second time.

```
newdata = circshift(data,-1);
data(isnan(newdata)==1)  = NaN;
data = data(1:2:end,:);
```

The shorelines can again be displayed by using

```
plot(data(:,1),data(:,2),'k'), axis equal
xlabel('Longitude'), ylabel('Latitude')
```

We then display the graphs in a similar layout to that used in the previous chapter. The corresponding MATLAB script reads as follows:

```
figure1 = figure('Color',[1 1 1]);
axes1 = axes('Units','Centimeters',...
    'Position',[3 2 10 8],...
    'Box','on',...
    'LineWidth',0.6,...
    'FontSize',8);
line1 = line(data(:,1),data(:,2),...
    'Color',[0.1 0.3 0.8],...
    'LineWidth',0.5);
title1 = title('GSHHS Shoreline Data Set',...
    'FontSize',10);
xlabel1 = xlabel('Longitude',...
    'FontSize',8);
ylabel1 = ylabel('Latitude',...
    'FontSize',8);
```

The length of the x-axis is 10 cm, the length of the y-axis is 8 cm, the line thickness of the axes is 0.6 pt (0.6 points), the line thickness of the shorelines is 0.5 pt, the axis labels and title are in a sans serif font such as *Helvetica*, the font size of the axis labels is 8 pt and that of the title is 10 pt (Fig. 6.1). The command

```
print -depsc2 coastline_linegraph_vs1_matlab.eps
```

creates an *Encapsulated Level 2 Color PostScript* (EPSC2) file, which is the preferred format to use for exporting vector graphics that contain line-drawings only, with no special effects such as interpolated shading or transparency.

Fig. 6.1 Display of the GSHHS shoreline data set. The map shows an area between latitudes 0° and 15° north, and longitudes 45° and 55° east. This simple map is made using the function plot, with equal axis aspect ratios.

6.3   The 2-Minute Gridded Global Relief Data ETOPO2

The *2-Minute Gridded Global Relief Data* (ETOPO2) is a global database of topography and bathymetry on a regular 2-minute grid (approximately 4 km). It is a compilation of data from a variety of sources. This data can be downloaded from the US National Geophysical Data Center (NGDC) web page

```
http://ngdc.noaa.gov/mgg/fliers/01mgg04.html
```

From the menu bar *Free On-line*, we select *custom grids*, which is linked to the *GEODAS Grid Translator*. First, we choose a Grid ID (e.g., *grid01*), the Grid Database (e.g., *ETOPO2 2-minute Global Relief Ver 2*), our computer system (e.g., *Macintosh*) and the Grid Format (e.g., *ASCII*) for both the data and the header. Second, we define the *latitude* and *longitude* bounds; for example, the latitude (*lat*) from 20 S to 20 N and a longitude (*lon*) between 30 E and 60 E covers the East African coast. The selected area can be transformed into a digital elevation matrix by pressing *Submit*. This matrix may be downloaded from the web page by pressing *Compress and Retrieve Your Grid* followed by *Retrieve compressed file* in the subsequent window. Decompressing the file *grid01.tgz* creates a directory *grid01_data*, which contains various data and help files. The subdirectory *grid01* contains the ASCII raster grid file *grid01.asc*, which has the following content:

```
NCOLS     901
NROWS    1201
XLLCORNER   30.00000
YLLCORNER  -20.00000
CELLSIZE 0.03333333
NODATA_VALUE   -32768
299     286     285     273     267     260 ...
298     279     282     273     263     254 ...
284     272     275     274     266     260 ...
267     267     269     270     272     269 ...
254     267     268     268     264     258 ...
(cont'd)
```

The headers document the size of the data matrix (901 columns and 1201 rows in our example), the coordinates of the lower-left corner ($x = 30$ and $y = -20$), the cell size ($0.033333 = 1/30$ degree latitude and longitude) and the -32768 flag for data voids. We comment the header by typing % at the beginning of the first six lines

```
%NCOLS     901
%NROWS    1201
%XLLCORNER   30.00000
%YLLCORNER  -20.00000
%CELLSIZE 0.03333333
%NODATA_VALUE   -32768
299     286     285     273     267     260 ...
298     279     282     273     263     254 ...
284     272     275     274     266     260 ...
267     267     269     270     272     269 ...
254     267     268     268     264     258 ...
(cont'd)
```

save the file as *etopo2_data.txt*, and then load the data into the workspace.

```
clear

ETOPO2 = load('etopo2_data.txt');
```

We flip the matrix up and down to change the indexing of the data array according to MATLAB conventions. The -32768 flag for data voids must then be replaced by the MATLAB representation for Not-a-Number NaN.

```
ETOPO2 = flipud(ETOPO2);
ETOPO2(find(ETOPO2 == -32768)) = NaN;
```

Finally, we check that the data are now correctly stored in the workspace by printing the minimum and maximum elevations for the area.

```
max(ETOPO2(:))
min(ETOPO2(:))
```

In this example, the maximum elevation for the area is 5,149 m and the minimum elevation is –5,656 m. The reference level is the sea level at 0 m. We now define a coordinate system using the information that the lower-left corner is at latitude 20° south and longitude 30° east. The resolution is 2 arc minutes, corresponding to 1/30 degree.

```
[LON,LAT] = meshgrid(30:1/30:60,-20:1/30:20);
```

We now generate a colored surface from the elevation data using the function `surf`.

```
surf(LON,LAT,ETOPO2)
shading interp
axis equal, view(0,90)
colorbar
```

This script opens a new figure window and generates a colored surface. The surface is highlighted by a set of color shades in an overhead view. We again display the graphs in a similar layout to that used in the previous chapter. The corresponding MATLAB script reads as follows:

```
figure1 = figure('Color',[1 1 1]);
axes1 = axes('Units','Centimeters',...
    'Position',[3 2 10 8],...
    'Box','on',...
    'LineWidth',0.6,...
    'FontSize',8);
surf1 = surf(LON,LAT,ETOPO2,...
    'Parent',axes1,...
    'FaceColor','interp',...
    'EdgeColor','none');
title1 = title('ETOPO2 Data Set',...
    'FontSize',10);
xlabel1 = xlabel('Longitude',...
    'FontSize',8);
ylabel1 = ylabel('Latitude',...
    'FontSize',8);
set(gca,'View',[0 90])
```

Details of the lines and fonts are as before. Furthermore, we set the axes property `view` to `[0,90]`. The command

VISUALIZING 3D DATA IN EARTH SCIENCES

6

```
print -depsc2 etopo2_pseudocolorplot_vs1_matlab.eps
```

exports the graph as an EPSC2 file. The result is obviously not a vector graph and we therefore export the graph as a JPEG file

```
print -djpeg70 -r600 etopo2_pseudocolorplot_vs2_matlab.jpg
```

compressed to 70 %, with a resolution of 600 dpi.

We next create a filled contour plot of the data. The contour intervals are

```
v = [-6000 -4000 -2000 -1000 -500 0 500 1000 2000 4000 6000];
```

Again, the length of the *x*-axis is 10 cm, the length of the *y*-axis is 8 cm, the line thickness of the axes is 0.6 pt (0.6 points), the line thickness of curves is 0.5 pt, the font of the axis labels and title is a sans serif font such as Helvetica, the font size of the axis labels is 8 pt, and that of the title is 10 pt (Fig. 6.2).

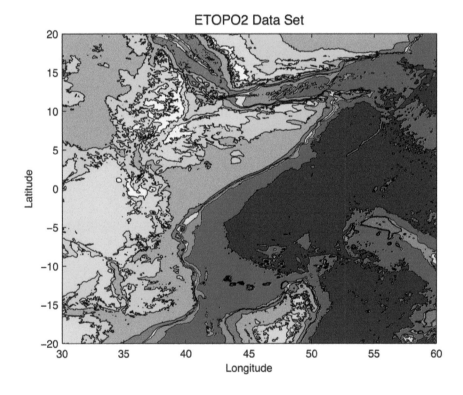

Fig. 6.2 Filled contour plot of the ETOPO2 elevation data. The function demcmap contained in the Mapping Toolbox creates and assigns a colormap appropriate for elevation data. According to the manual of the toolbox, the number of land and sea colors in the colormap is in proportion to the maximum elevations and depths in the matrix map.

```
figure1 = figure('Color',[1 1 1]);
axes1 = axes('Units','Centimeters',...
    'Position',[3 2 10 8],...
    'Box','on',...
    'LineWidth',0.6,...
    'FontSize',8);
contourf1 = contourf(LON,LAT,ETOPO2,v)
title1 = title('ETOPO2 Data Set',...
    'FontSize',10);
xlabel1 = xlabel('Longitude',...
    'FontSize',8);
ylabel1 = ylabel('Latitude',...
    'FontSize',8);
set(gcf,'Colormap',demcmap(ETOPO2))
```

The function `demcmap` contained in the Mapping Toolbox creates and assigns a colormap appropriate for elevation data. According to the toolbox manual the number of land and sea colors in the colormap is in proportion to the maximum elevations and depths in the matrix map. We can export the graph as an EPSC2 file:

```
print -depsc2 etopo2_filledcontourplot_vs3_matlab.eps
```

We can change the view of a surface plot using `view`.

```
figure1 = figure('Color',[1 1 1]);
axes1 = axes('Visible','off',...
    'Units','centimeters',...
    'Position',[3 2 10 8]);
hold(axes1,'all');
surf1 = surf(LON,LAT,ETOPO2,...
    'FaceColor','interp',...
    'EdgeColor','none');
set(gca,'View',[20 60])
set(gcf,'Colormap',demcmap(ETOPO2))
```

The command

```
print -depsc2 etopo2_surfaceplot_vs1_matlab.eps
```

exports the graph as an EPSC2 file. Once again, the result is obviously not a vector graph and we therefore export the graph as a JPEG file

```
print -djpeg70 -r600 etopo2_surfaceplot_vs2_matlab.jpg
```

compressed to 70 %, with a resolution of 600 dpi. We can then combine the 3D graph with 3D contours stored in v (Fig. 6.3).

```
figure1 = figure('Color',[1 1 1]);
axes1 = axes('Visible','off',...
    'Units','centimeters',...
```

Fig. 6.3 Surface plot of the ETOPO2 elevation data, in combination with 3D contours. The function demcmap contained in the Mapping Toolbox creates and assigns a colormap appropriate for elevation data. The function view uses an azimuth of 20° and an elevation of 60° to change the view an observer sees of the 3D plot.

```
    'Position',[3 2 10 8],...
    'FontSize',8);
hold(axes1,'all');
surf1 = surf(LON,LAT,ETOPO2,...
    'FaceColor','interp',...
    'EdgeColor','none');
contour31 = contour3(LON,LAT,ETOPO2,v,'k');
set(gca,'View',[20 60])
set(gcf,'Colormap',demcmap(ETOPO2))
```

As before, the command

```
print -depsc2 etopo2_surfaceplotcontours_vs1_matlab.eps
```

exports the graph as an EPSC2 file. Yet again, the result is obviously not a vector graph and we therefore export the graph as a JPEG file

```
print -djpeg70 -r600 etopo2_surfaceplotcontours_vs2_matlab.jpg
```

compressed to 70 %, with a resolution of 600 dpi. A spectacular way to plot digital topography is as a surface plot with angled lighting (Fig. 6.4).

Fig. 6.4 Surface plot of the ETOPO2 elevation data using light. The plot uses *Phong* as the lighting type, which is very popular in 3D computer graphics, creating a combined diffuse and specular reflection on surfaces. The *specular exponent* of 20 defines the size of the highlight spot on the surface, which can vary between 0 and 500.

```
figure1 = figure('Color',[1 1 1]);
axes1 = axes('Visible','off',...
     'Units','centimeters',...
     'Position',[3 2 10 8]);
hold(axes1,'all');
surf1 = surf(LON,LAT,ETOPO2,...
     'SpecularExponent',20,...
     'FaceLighting','phong',...
     'FaceColor','interp',...
     'EdgeColor','none');
light1 = light('Parent',axes1,...
     'Style','local',...
     'Position',[145 70 155801]);
set(gca,'View',[20 60])
```

We again export the graph as a JPEG file

```
print -djpeg70 -r600 etopo2_surfaceplotlight_vs1_matlab.jpg
```

compressed to 70 %, with a resolution of 600 dpi.

6.4   The Global 30-Arc Second Elevation Data GTOPO30

The *Global 30 Arc Second Elevation Data* (GTOPO30) is a 30 arc second (approximately 1 km) global digital elevation data set that contains only elevation data, not bathymetry. The data set has been developed by the Earth Resources Observation System Data Center and is available from the web page

```
http://eros.usgs.gov/#/Find_Data/Products_and_Data_Available/
gtopo30_info
```

The data set draws on a variety of international data sources, but is mainly based on raster data from the *Digital Terrain Elevation Model* (DTEM) and vector data from the *Digital Chart of the World* (DCW). The GTOPO30 data set has been divided into 33 tiles. The tile names refer to the longitude and latitude of the upper-left (northwest) corner of the tile (in contrast to the lower left corner used in Section 6.3 and again in Section 6.5). Thus the name *e020n40* refers to the tile that has coordinates for its upper-left corner of longitude 20 degrees east and latitude 40 degrees north. As an example, we select and download the tile *e020n40*, which is provided as a 24.9 MB compressed *tar* file named *e020n40.tar.gz*. After decompressing the file, we obtain eight files containing the raw data and header files in various formats. The *tar* file also provides a GIF image of a shaded relief display of the data.

Importing the GTOPO30 data into the workspace is simple. The Mapping Toolbox provides an import routine `gtopo30` with which to read the data contained in the two files *E020N40.DEM* and *E020N40.HRD* that provide the actual digital topography and a header, and to store it on a regular data grid. We import only a subset of the original matrix:

```
clear

latlim = [-5 5]; lonlim = [30 40];
GTOPO30 = gtopo30('E020N40',1,latlim,lonlim);
```

This script reads the data from the tile *e020n40* (without file extension) at full resolution (scale factor = 1) into the matrix GTOPO30, which has the dimension of $1,200 \times 1,200$ elements. The coordinate system is defined by using the *lon/lat* limits listed above. The resolution is 30 arc seconds, corresponding to 1/120 degrees.

```
[LON,LAT] = meshgrid(30:1/120:40-1/120,-5:1/120:5-1/120);
```

We need to reduce the limits by 1/120 to obtain a matrix of similar dimensions to the GTOPO30 matrix. A grayscale image can be generated from the elevation data using the function surf. The colormap is flipped vertically in order to obtain dark colors for high elevations and light colors for low elevations. The fourth power of the colormap gray is then used to darken the higher elevations on the plot even more.

```
surf(LON,LAT,GTOPO30)
shading interp
colormap(flipud(gray.^4))
axis equal, view(0,90)
colorbar
```

This script opens a new figure window and generates the gray surface using interpolated shading in an overhead view (Fig. 6.3). We again display the graphs in a similar layout to that used in the previous chapter. The corresponding MATLAB script reads as follows:

```
v = 0 : 500 : 6000;
figure1 = figure('Color',[1 1 1]);
axes1 = axes('Units','Centimeters',...
    'Position',[3 2 10 8],...
    'Box','on',...
    'LineWidth',0.6,...
    'FontSize',8);
contourf1 = contourf(LON,LAT,GTOPO30,v);
title1 = title('GTOPO30 Data Set',...
    'FontSize',10);
xlabel1 = xlabel('Longitude',...
    'FontSize',8);
ylabel1 = ylabel('Latitude',...
    'FontSize',8);
demcmap(GTOPO30)
```

The length of the *x*-axis is 10 cm, the length of the *y*-axis is 8 cm, the line thickness of the axes is 0.6 pt (0.6 points), the axis labels and title are in a sans serif font such as *Helvetica*, the font size of the axis labels is 8 pt and that of the title is 10 pt. The function demcmap contained in the Mapping Toolbox again creates and assigns a colormap appropriate for elevation data. The command

```
print -depsc2 gtopo30_filledcontourplot_vs1_matlab.eps
```

exports the graph as an EPSC2 file. We again plot the digital topography as a surface plot with angled lighting (Fig. 6.5).

```
figure1 = figure('Color',[1 1 1]);
axes1 = axes('Visible','off',...
    'Units','centimeters',...
    'Position',[3 2 10 8]);
hold(axes1,'all');
surf1 = surf(LON,LAT,GTOPO30,...
```

Fig. 6.5 Surface plot of the GTOPO30 elevation data using `light`. The plot uses similar lighting settings to the previous plot, including the *Phong* lighting type and a *specular exponent* of 20.

```
    'SpecularExponent',20,...
    'FaceLighting','phong',...
    'FaceColor','interp',...
    'EdgeColor','none');
light1 = light('Parent',axes1,...
    'Style','local',...
    'Position',[145 70 155801]);
set(gca,'View',[25 20])
colormap hsv
daspect([1 1 4000])
```

We again export the graph as a JPEG file

```
print -djpeg70 -r600 gtopo30_surfaceplotlight_vs1_matlab.jpg
```

compressed to 70 %, with a resolution of 600 dpi.

6.5 The Shuttle Radar Topography Mission SRTM

The *Shuttle Radar Topography Mission* (SRTM) consisted of a radar system that flew onboard the Space Shuttle *Endeavour* during an 11-day mission in February 2000. SRTM was an international project spearheaded by the *National Geospatial-Intelligence Agency* (NGA) and the National *Aeronautics and Space Administration* (NASA). Detailed information on the SRTM project, including a gallery of images and a user's forum, can be accessed through the NASA web page:

```
http://www2.jpl.nasa.gov/srtm/
```

The data were processed at the *Jet Propulsion Laboratory* of the *California Institute of Technology* (Caltech). They are distributed through the *EROS*

Data Center of the *United States Geological Survey* (USGS), using the Seamless Data Distribution System.

```
http://seamless.usgs.gov/
```

Alternatively, the raw data files can be downloaded from

```
http://dds.cr.usgs.gov/srtm/
```

This directory contains zipped files of SRTM *digital elevation models* (DEMs) from various areas of the world, processed by the SRTM global processor, and sampled at resolutions of 1 arc second (SRTM1, 30 meter grid), 3 arc seconds (SRTM3, 90 meter grid) and 30 arc seconds (SRTM30, about 1 km grid, to commensurate with GTOPO30). As an example, we download the 1.7 MB file *s01e036.hgt.zip* from

```
http://dds.cr.usgs.gov/srtm/version2_1/SRTM3/Africa/
```

The file contains SRTM3 data for the Kenya Rift Valley in East Africa. All elevations are in meters, referenced to the WGS84 EGM96 geoid as documented at

```
http://earth-info.nga.mil/GandG/wgs84/index.html
```

The name of this file refers to the longitude and latitude of the lower-left (southwest) pixel of the tile, i.e., latitude one degree south and longitude 36 degrees east. SRTM3 data contain 1,201 lines and 1,201 samples, with similar numbers of overlapping rows and columns. After having downloaded and unzipped the file, we save *s01e036.hgt* in our working directory and rename it *srtm_data.hgt*. The digital elevation model is provided as 16-bit signed integer data in a simple binary raster. Bit order is Motorola (*big-endian*) standard with the most significant bit first. The data are imported into the workspace using

```
clear

fid = fopen('srtm_data.hgt','r');
SRTM = fread(fid,[1201,inf],'int16','b');
fclose(fid);
```

This script opens the file *srtm_data.hgt* for read-only access using `fopen` and defines the file identifier `fid`, which is then used by `fread` for reading the binaries from the file, and writing them into the matrix `SRTM`. Function `fclose` closes the file defined by `fid`. The matrix needs to be transposed and flipped vertically.

```
SRTM = SRTM'; SRTM = flipud(SRTM);
```

The –*32768* flag for data voids can be replaced by NaN, which is the MATLAB representation for Not-a-Number.

```
SRTM(find(SRTM == -32768)) = NaN;
```

The SRTM data contain gaps that might cause spurious effects during statistical analysis or when displaying the digital elevation model in a graph. A popular way to eliminate gaps in digital elevation models is to fill any such gaps with the arithmetic means of adjacent elements. We use the function nanmean since it treats NaNs as missing values and returns the mean of the remaining elements that are not NaNs. The following double for loop averages SRTM(i-1:i+1,j-1:j+1) arrays, i.e., averages over three-by-three element wide areas of the digital elevation model.

```
for i = 2 : 1200
    for j = 2 : 1200
        if isnan(SRTM(i,j)) == 1
            SRTM(i,j) = nanmean(nanmean(SRTM(i-1:i+1,j-1:j+1)));
        end
    end
end
clear i j
```

Finally, we check that the data are now correctly stored in the workspace by printing the minimum and maximum elevations of the area.

```
max(SRTM(:))

ans =
   3992

min(SRTM(:))

ans =
   1504
```

In our example, the maximum elevation of the area is 3,992 m above sea level and the minimum is 1,504 m. A coordinate system can be defined by using the information that the lower-left corner is *s01e036*. The resolution is 3 arc seconds, corresponding to 1/1,200 degree.

```
[LON,LAT] = meshgrid(36:1/1200:37,-1:1/1200:0);
```

A shaded grayscale map can be generated from the elevation data using the function surfl. This function displays a shaded surface with simulated lighting.

```
surfl(LON,LAT,SRTM)
shading interp
colormap gray
view(0,90)
```

This script opens a new figure window and generates the shaded-relief map using interpolated shading, as well as a gray colormap in an overhead view. Since SRTM data contain a large amount of noise, we first smooth the data using an arbitrary 9×9 pixel moving average filter. The new matrix is then stored in the matrix SRTM_FILTERED.

```
B = 1/81 * ones(9,9);
SRTM_FILTERED = filter2(B,SRTM);
```

The corresponding shaded-relief map is generated by

```
surfl(LON,LAT,SRTM_FILTERED)
shading interp
colormap gray
view(0,90)
```

After having generated the shaded-relief map (Fig. 6.4), the plot must be exported to a graphics file. For instance, the figure may be written into a JPEG format with a 70 % quality level and 300 dpi resolution.

```
print -djpeg70 -r300 srtm_surfaceplotlight_vsl_matlab.jpg
```

The new file *srtmimage.jpg* has a size of 320 KB; the decompressed image has a size of 16.5 MB.

We again display the graphs in a similar layout to that used in the previous chapter. The corresponding MATLAB script reads as follows:

```
figure1 = figure('Color',[1 1 1]);
axes1 = axes('Units','Centimeters',...
    'Position',[3 2 10 8],...
    'Box','on',...
    'LineWidth',0.6,...
    'FontSize',8);
surfl1 = surfl(LON,LAT,SRTM_FILTERED);
title1 = title('SRTM Data Set',...
    'FontSize',10);
xlabel1 = xlabel('Longitude',...
    'FontSize',8);
ylabel1 = ylabel('Latitude',...
    'FontSize',8);
shading interp
colormap gray
view(0,90)
```

The length of the x-axis is 10 cm, the length of the y-axis is 8 cm, the line thickness of the axes is 0.6 pt (0.6 points), the axis labels and title are in a sans serif font such as *Helvetica*, the font size of the axis labels is 8 pt and that of the title is 10 pt. We export the graph using the command

```
print -djpeg70 -r600 srtm_surfaceplotlight_vs2_matlab.jpg
```

in a JPEG format with a 70 % quality level and 300 dpi resolution. We again plot the topography as a surface plot with angled light (Fig. 6.6).

```
figure1 = figure('Color',[1 1 1]);
axes1 = axes('Visible','off',...
    'Units','centimeters',...
    'Position',[3 2 10 8],...
    'FontSize',8);
hold(axes1,'all');
surf1 = surf(LON,LAT,SRTM_FILTERED,...
    'SpecularExponent',20,...
    'FaceLighting','phong',...
    'FaceColor','interp',...
    'EdgeColor','none');
light1 = light('Parent',axes1,...
    'Style','local',...
    'Position',[145 70 155801]);
set(gca,'View',[25 20])
daspect([1 1 12000])
```

We again export the graph as a JPEG file

```
print -djpeg70 -r600 srtm_surfaceplotlight_vs3_matlab.jpg
```

compressed to 70 %, with a resolution of 600 dpi.

Fig. 6.6 Surface plot of the SRTM elevation data using `light`. The plot uses similar lighting settings to the previous plot, including the *Phong* lighting type and a *specular exponent* of 20.

6.6 Interpolating and Visualizing Irregularly-Spaced Data

The previous data sets were all stored in evenly-spaced two-dimensional arrays. Most ground-based earth science data sets are, however, obtained from irregular sampling patterns. The data are therefore unevenly-spaced and need to be interpolated in order to allow a smooth and continuous surface to be computed from field measurements. MATLAB has, from the start, provided a biharmonic spline interpolation method that was developed by Sandwell (1987). This gridding method is particularly well suited for producing smooth surfaces from noisy data sets with unevenly-distributed control points. MATLAB today also provides other interpolation techniques, such as bilinear, cubic spline, and nearest neighbor interpolation.

As an example, we use synthetic *xyz* data representing the vertical distance between the surface of an imaginary stratigraphic horizon that has been displaced by a normal fault, and a reference surface. The footwall of the fault contains roughly horizontal strata, whereas the hanging wall comprises two large sedimentary basins. The *xyz* data are irregularly distributed and so need to be interpolated onto a regular grid. The data are stored as a three-column table in a file named *normalfault.txt*.

```
4.3229698e+02    7.4641694e+01    9.7283620e-01
4.4610209e+02    7.2198697e+01    6.0655065e-01
4.5190255e+02    7.8713355e+01    1.4741054e+00
4.6617169e+02    8.7182410e+01    2.2842172e+00
4.6524362e+02    9.7361564e+01    1.1295175e-01
4.5526682e+02    1.1454397e+02    1.9007110e+00
4.2930233e+02    7.3175896e+01    3.3647807e+00
(cont'd)
```

The first and second columns contain the coordinates x (between 420 and 470) and y (between 70 and 120) of an arbitrary spatial coordinate system, while the third column contains the vertical z values. The data are loaded using

```
clear

data = load('normalfault_data.txt');
```

We wish to first create an overview plot from the spatial distribution of the control points. In order to label the points in the plot, numerical z values from the third column are converted into character string representations with a maximum of two digits.

```
labels = num2str(data(:,3),2);
```

The 2D plot of our data is generated in two steps. Firstly, the data are displayed as empty circles using the `plot` command. Secondly, the data are labeled by using the function `text(x,y,'string')`, which adds text, contained in a character array `labels`, to the *xy* locations. The value 1 is added to all *x* coordinates in order to produce a small offset between the circles and the text.

```
plot(data(:,1),data(:,2),'o'), hold on
text(data(:,1)+1,data(:,2),labels), hold off
```

This plot helps us to define the axis limits for gridding and contouring, `xlim = [420 470]` and `ylim = [70 120]`. The function `meshgrid` transforms the domain specified by vectors *x* and *y* into arrays `XI` and `YI`. The rows of the output array `XI` are copies of the vector x and the columns of the output array `YI` are copies of the vector y. We choose 1.0 as the grid interval.

```
x = 420:1:470; y = 70:1:120;
[XI,YI] = meshgrid(x,y);
```

The biharmonic spline interpolation is used to interpolate the irregularly-spaced data at the grid points specified by `XI` and `YI`.

```
ZI = griddata(data(:,1),data(:,2),data(:,3),XI,YI,'v4');
```

The option `v4` selects the biharmonic spline interpolation. MATLAB provides various tools with which to display the results. The simplest way to display the gridding results is as a contour plot using `contour`. The number of contour levels and the values of the contour levels are chosen automatically, by default. The choice of the contour levels depends on the minimum and maximum values of *z*.

```
contour(XI,YI,ZI)
```

Alternatively, the number of contours can be chosen manually, e.g., ten contour levels can be chosen as follows:

```
contour(XI,YI,ZI,10)
```

Contouring can also be performed at values specified in a vector v. Since the maximum and minimum values of *z* are

```
min(data(:,3))

ans =
   -27.4357

max(data(:,3))
```

```
ans =
    21.3018
```

we choose

```
v = -40 : 10 : 20;
```

The command

```
[c,h] = contour(XI,YI,ZI,v);
```

yields contour matrix c and a handle h that can be used as input to the function clabel, which labels contours automatically.

```
clabel(c,h)
```

Alternatively, the plot can be labeled manually by selecting the manual option in the function clabel. This function places labels onto locations that have been selected with the mouse. Labeling is terminated by pressing the return key.

```
[c,h] = contour(XI,YI,ZI,v);
clabel(c,h,'manual')
```

Exporting contour plots often results in broken polygons. A possible way around this is to create filled contours, export the graph as an EPS file, and then unfill the polygons in the vector graphics software. The function contourf is used to create the gridded data, after which the function colorbar is used to create a legend for the plot. We can also plot the locations (small circles) and z values (contour labels) of the true data points (Fig. 6.7).

```
contourf(XI,YI,ZI,v), colorbar, hold on
plot(data(:,1),data(:,2),'ko')
text(data(:,1)+1,data(:,2),labels), hold off
```

A pseudocolor plot is generated by using the function pcolor. Black contours are also added at the previously defined levels v defined before.

```
pcolor(XI,YI,ZI), shading flat, hold on
contour(XI,YI,ZI,v,'k'), hold off
```

The third dimension is added to the plot using the mesh command. We can also use this example to introduce the function view(az,el) to specify the direction of viewing. In this function, az is the azimuth or horizontal rotation and el is the elevation (both in degrees). The values $az = -37.5$ and $el = 30$ define the default view for all 3D plots,

Fig. 6.7 Contour plot with the locations (small circles) and *z*-values (contour labels) of the true data points.

```
mesh(XI,YI,ZI), view(-37.5,30)
```

whereas $az = 0$ and $el = 90$ is directly overhead and the default 2D view

```
mesh(XI,YI,ZI), view(0,90)
```

The function mesh is one of many methods available in MATLAB for 3D presentation, another commonly used function being surf. The figure may be rotated by selecting the *Rotate 3D* option on the *Edit Tools* menu. We also introduce the function colormap, which uses predefined color look-up tables for 3D graphs. Typing help graph3d lists a number of built-in colormaps, although colormaps can also be arbitrarily modified and generated by the user. As an example, we use the colormap hot, which is a black-red-yellow-white colormap.

```
surf(XI,YI,ZI), colormap('hot'), colorbar
```

Using *Rotate 3D* only rotates the 3D plot, not the colorbar. The function `surfc` combines both a surface and 2D contours in a single graph.

```
surfc(XI,YI,ZI)
```

The function `surfl` can be used to illustrate an advanced application for 3D visualization, generating a 3D colored surface with interpolated shading and lighting. The axis labeling, ticks, and background can be turned off by typing `axis off`. In addition, black 3D contours can be added to the surface at the previously defined intervals v. The grid resolution is increased prior to data plotting in order to obtain smooth surfaces (Fig. 6.8).

```
[XI,YI] = meshgrid(420:0.25:470,70:0.25:120);
ZI = griddata(data(:,1),data(:,2),data(:,3),XI,YI,'v4');

surf(XI,YI,ZI), shading interp, light, axis off, hold on
contour3(XI,YI,ZI,v,'k'), hold off
```

We again display the graphs in a similar layout to that used in the previous chapter. The corresponding MATLAB script reads as follows:

```
figure1 = figure('Color',[1 1 1]);
axes1 = axes('Units','Centimeters',...
    'Position',[3 2 10 8]);
surf1 = surf(XI,YI,ZI);
shading interp
light
```

Fig. 6.8 Three-dimensional surface with interpolated shading and simulated lighting. The axis labeling, ticks, and background are all turned off. The plot shows 3D contours as black lines.

VISUALIZING 3D DATA IN EARTH SCIENCES

6

```
axis off
hold on
contour31 = contour3(XI,YI,ZI,v,'k');
hold off
```

The length of the hidden x-axis is 10 cm and the length of the hidden y-axis is 8 cm. We export the graph using the command

```
print -djpeg70 -r600 normalfault_surfaceplotlight_vs1_matlab.jpg
```

in a JPEG format with a 70 % quality level and 300 dpi resolution.

Recommended Reading

Sandwell DT (1987) Biharmonic Spline Interpolation of GEOS-3 and SEASAT Altimeter data. Geophysical Research Letters 2:139–142
Trauth MH (2010) MATLAB Recipes for Earth Sciences – Third Edition. Springer, Berlin Heidelberg New York

7 Processing and Displaying Images in Earth Sciences

7.1 Introduction

Computer graphics are stored and processed as either vector or raster data. Most of the data types that were encountered in the previous chapter were vector data, i.e., points, lines and polygons. Images are generally presented as raster data, i.e., as a 2D array of color intensities. Images are everywhere in geosciences. Field geologists use aerial photographs and satellite images to identify lithologic units, tectonic structures, landslides and other features within a study area. Geomorphologists use such images for the analysis of drainage networks, river catchments, and vegetation or soil types. The analysis of images from thin sections, the automated identification of objects, and the measurement of varve thicknesses all make use of a great variety of image processing methods.

This chapter is concerned with the analysis and display of image data. The various ways that raster data can be stored on a computer are explained in Section 7.2. The main tools for importing, manipulating and exporting image data are presented in Section 7.3. This information is then used to process and to georeference satellite images (Sections 7.4 and 7.5). On-screen digitization techniques are discussed in Section 7.6. While the MATLAB User's Guide to the Image Processing Toolbox provides an excellent general introduction to the analysis of images, this chapter provides an overview of typical applications in the earth sciences.

7.2 Storing Images on a Computer

Vector and raster graphics are the two fundamental methods for storing images. The typical format for storing *vector data* has already been introduced in the previous chapters. In the following example, the two columns in the file *coastline.txt* represent the longitudes and latitudes of the points of a polygon.

M. H. Trauth and E. Sillmann, *MATLAB® and Design Recipes for Earth Sciences*, **149**
DOI: 10.1007/978-3-642-32544-1_7, © Springer-Verlag Berlin Heidelberg 2013

```
NaN              NaN
42.892067 0.000000
42.893692 0.001760
NaN              NaN
42.891052 0.001467
42.898093 0.007921
42.904546 0.013201
42.907480 0.016721
42.910414 0.020828
42.913054 0.024642
(cont'd)
```

The NaNs help to identify break points in the data (Section 6.2).

In contrast, *raster data* are stored as 2D arrays. The elements of these arrays represent variables such as the altitude of a grid point above sea level, the annual rainfall or, in the case of an image, the color intensity values.

```
174 177 180 182 182 182
165 169 170 168 168 170
171 174 173 168 167 170
184 186 183 177 174 176
191 192 190 185 181 181
189 190 190 188 186 183
```

Raster data can be visualized as a 3D plot. The x and y are the indices of the 2D array or any other reference frame, and z is the numerical value of the elements of the array (see also Chapter 4). Alternatively, the numerical values contained in the 2D array can be displayed as a pseudocolor plot, which is a rectangular array of cells with colors determined by a colormap. A colormap is an m-by-3 array of real numbers between 0.0 and 1.0. Each row defines a red, green, blue (RGB) color. An example is the above array that could be interpreted as grayscale intensities ranging from 0 (black) to 255 (white). More complex examples include satellite images that are stored in 3D arrays.

As previously discussed, a computer stores data as bits that have one of two states, one and zero (Chapter 4). If the elements of the 2D array represent the color intensity values of the *pixels* (short for *picture elements*) of an image, 1-bit arrays contain only ones and zeros.

```
0   0   1   1   1   1
1   1   0   0   1   1
1   1   1   1   0   0
1   1   1   1   0   1
0   0   0   0   0   0
0   0   0   0   0   0
```

This 2D array of ones and zeros can be simply interpreted as a black-and-white image, where the value of one represents white and zero corresponds to black. Alternatively, the 1-bit array could be used to store an image consisting of any two different colors, such as red and blue.

In order to store more complex types of data, the bits are joined together to form larger groups, such as bytes consisting of eight bits. Since the earliest computers could only process eight bits at a time, early computer code was written in sets of eight bits, which came to be called bytes. Hence, each element of the 2D array or pixel contains a vector of eight ones or zeros.

```
1      0      1      0      0      0      0      1
```

These 8 bits (or 1 byte) allow $2^8 = 256$ possible combinations of the eight ones or zeros. 8 bits are therefore able to represent 256 different intensities such as grayscales. The 8 bits can be read in the following way reading from right to left: a single bit represents two numbers, two bits give four numbers, three bits show eight numbers, and so forth up to a byte, or eight bits, which represents 256 numbers. Each added bit doubles the count of numbers. Here is a comparison of binary and decimal representations of the number 161:

```
128    64    32    16    8     4     2     1         (value of the bit)
 1      0     1     0    0     0     0     1         (binary)

128 +   0 + 32  + 0 +   0 +   0 +   0 +   1 = 161    (decimal)
```

The end members of the binary representation of grayscales are

```
0      0      0      0      0      0      0      0
```

which is black, and

```
1      1      1      1      1      1      1      1
```

which is pure white. In contrast to the above 1-bit array, the one-byte array allows a grayscale image of 256 different levels to be stored. Alternatively, the 256 numbers could be interpreted as 256 discrete colors. In either case, the display of such an image requires an additional source of information concerning how the 256 intensity values are converted into colors. Numerous global colormaps for the interpretation of 8-bit color images exist that allow the cross-platform exchange of raster images, while local colormaps are often embedded in a graphics file.

The disadvantage of 8-bit color images is that the 256 discrete colorsteps are not enough to simulate smooth transitions for the human eye. A 24-bit system is therefore used in many applications, with 8 bits of data for each RGB channel giving a total of $256^3 = 16,777,216$ colors. Such a 24-bit image

7 | PROCESSING AND DISPLAYING IMAGES IN EARTH SCIENCES

is stored in three 2D arrays, or one 3D array, of intensity values between 0 and 255.

```
195   189   203   217   217   221
218   209   187   192   204   206
207   219   212   198   188   190
203   205   202   202   191   201
190   192   193   191   184   190
186   179   178   182   180   169

209   203   217   232   232   236
234   225   203   208   220   220
224   235   229   214   204   205
223   222   222   219   208   216
209   212   213   211   203   206
206   199   199   203   201   187

174   168   182   199   199   203
198   189   167   172   184   185
188   199   193   178   168   172
186   186   185   183   174   185
177   177   178   176   171   177
179   171   168   170   170   163
```

Compared to the 1-bit and 8-bit representations of raster data, the 24-bit storage certainly requires a lot more computer memory. In the case of very large data sets such as satellite images and digital elevation models the user should therefore think carefully about the most suitable way to store the data. The default data type in MATLAB is the 64-bit array, which allows storage of the sign of a number (first bit), the exponent (bits 2 to 12) and roughly 16 significant decimal digits in the range of approximately 10^{-308} to 10^{+308} (bits 13 to 64). However, MATLAB also works with other data types, such as 1-bit, 8-bit and 24-bit raster data, to save memory.

The memory required for storing a raster image depends on the data type and the image's dimension. The dimension of an image can be described by the number of pixels, which is the number of rows multiplied by the number of columns of the 2D array. Let us assume an image of 729×713 pixels, such as the one we will use in the following section. If each pixel needs 8 bits to store a grayscale value, the memory required by the data is $729 \times 713 \times 8 = 4{,}158{,}216$ bits or $4{,}158{,}216/8 = 519{,}777$ bytes. This number is exactly what we obtain by typing whos in the command window. Common prefixes for bytes are kilo-, mega-, giga- and so forth.

```
bit = 1 or 0  (b)
8 bits = 1 byte  (B)
1024 bytes = 1 kilobyte  (KB)
1024 kilobytes = 1 megabyte  (MB)
1024 megabytes = 1 gigabyte  (GB)
1024 gigabytes = 1 terabyte  (TB)
```

Note that in data communication 1 kilobit = 1,000 bits, while in data storage 1 kilobyte = 1,024 bytes. A 24-bit or *true color image* then requires three times the memory required to store an 8-bit image, or 1,559,331 bytes = 1,559,331/1,024 kilobytes (kB) ≈ 1,523 kB ≈ 1,559,331/1,024^2=1.487 megabytes (MB).

However, the dimensions of an image are often given, not by the total number of pixels, but by its length, height, and resolution. The resolution of an image is the number of *pixels per inch* (ppi) or *dots per inch* (dpi). The standard resolution of a computer monitor is 72 dpi although modern monitors often have a higher resolution such as 96 dpi. For instance, a 17 inch monitor with 72 dpi resolution displays 1,024 × 768 pixels. If the monitor is used to display images at a different (lower, higher) resolution, the image is resampled to match the monitor's resolution. For scanning and printing, a resolution of 300 or 600 dpi is enough in most applications. However, scanned images are often scaled for large printouts and therefore have higher resolutions such as 2,400 dpi. The image used in the next section has a width of 25.2 cm (or 9.92 inches) and a height of 25.7 cm (10.12 inches). The resolution of the image is 72 dpi. The total number of pixels is therefore 72 × 9.92 ≈ 713 in a horizontal direction, and 72 × 10.12 ≈ 729 in a vertical direction.

7.3 Importing, Processing and Exporting Images

We first need to learn how to read an image from a graphics file into the workspace. As an example, we use a satellite image showing a 10.5 km by 11 km subarea in northern Chile:

```
http://asterweb.jpl.nasa.gov/gallery/images/unconform.jpg
```

The file *unconform.jpg* is a processed TERRA-ASTER satellite image that can be downloaded free-of-charge from the NASA web page. We save this image in the working directory as *unconform_image_vs1_original.jpg*. The command

```
clear

unconform1 = imread('unconform_image_vs1_original.jpg');
```

reads and decompresses the JPEG file, imports the data as a 24-bit RGB image array and stores the data in a variable unconform1. The command

```
whos
```

shows how the RGB array is stored in the workspace:

```
Name                    Size                    Bytes  Class     Attributes
unconform1              729x713x3               1559331  uint8
```

The details indicate that the image is stored as a $729 \times 713 \times 3$ array representing a 729×713 array for each of the colors red, green and blue. The listing of the current variables in the workspace also gives the information *uint8* array, i.e., each array element representing one pixel contains 8-bit integers. These integers represent intensity values between 0 (minimum intensity) and 255 (maximum). As an example, here is a sector in the upper-left corner of the data array for red:

```
unconform1(50:55,50:55,1)

ans =
    174 177 180 182 182 182
    165 169 170 168 168 170
    171 174 173 168 167 170
    184 186 183 177 174 176
    191 192 190 185 181 181
    189 190 190 188 186 183
```

Next, we can view the image using the command

```
imshow(unconform1)
```

which opens a new Figure Window showing an RGB composite of the image.

In contrast to the RGB image, a grayscale image needs only a single array to store all the necessary information. We therefore convert the RGB image into a grayscale image using the command `rgb2gray` (RGB to gray):

```
unconform2 = rgb2gray(unconform1);
```

The new workspace listing now reads

```
Name                    Size                    Bytes  Class     Attributes
ans                     6x6                        36  uint8
unconform1              729x713x3               1559331  uint8
unconform2              729x713                 519777  uint8
```

in which the difference between the 24-bit RGB and the 8-bit grayscale arrays can be observed. The command

```
imshow(unconform2)
```

displays the result. It is easy to see the difference between the two images in separate Figure Windows (Fig 7.1). Let us now process the grayscale image. First, we compute a histogram of the distribution of intensity values.

```
imhist(unconform2)
```

A simple technique to enhance the contrast of such an image is to transform this histogram to obtain an equal distribution of grayscales.

```
unconform3 = histeq(unconform2);
```

PROCESSING AND DISPLAYING IMAGES IN EARTH SCIENCES

7

Fig. 7.1 Grayscale image. After converting the RGB image stored in a $729 \times 713 \times 3$ array into a grayscale image stored in a 729×713 array, the result is displayed using imshow. Original image courtesy of NASA/GSFC/METI/ERSDAC/JAROS and U.S./Japan ASTER Science Team.

We can view the difference again using

```
imshow(unconform3)
```

and save the results in a new file.

```
imwrite(unconform3,'unconform_image_vs2_matlab.jpg')
```

We can read the header of the new file by typing

```
imfinfo('unconform_image_vs2_matlab.jpg')
```

which yields

```
        Filename:  'unconform_image_vs2_matlab.jpg'
     FileModDate:  '26-Oct-2011 22:54:36'
        FileSize:  138419
          Format:  'jpg'
   FormatVersion:  ''
           Width:  713
          Height:  729
        BitDepth:  8
       ColorType:  'grayscale'
 FormatSignature:  ''
 NumberOfSamples:  1
    CodingMethod:  'Huffman'
   CodingProcess:  'Sequential'
         Comment:  {}
```

Hence, the command `imfinfo` can be used to obtain useful information (name, size, format and color type) concerning the newly-created image file.

There are many ways of transforming the original satellite image into a practical file format. The image data could, for instance, be stored as an *indexed color image*, which consists of two parts: a colormap array and a data array. The colormap array is an m-by-3 array containing floating-point values between 0 and 1. Each column specifies the intensity of the red, green and blue colors. The data array is an x-by-y array containing integer elements corresponding to the lines m of the colormap array, i.e., the specific RGB representation of a certain color. Let us transfer the above RGB image into an indexed image. The colormap of the image should contain 16 different colors. The result of

```
[x,map] = rgb2ind(unconform1,16);
imshow(unconform1), figure, imshow(x,map)
```

clearly shows the difference between the original 24-bit RGB image (256^3 or about 16.7 million different colors) and a color image of only 16 different colors.

7.4   Processing and Printing Satellite Images

In the previous section, we used a processed ASTER image that we down-loaded from the ASTER web page. The original ASTER raw data contain a lot more information and higher resolution than the free-of-charge image stored in *unconform.jpg*. The ASTER instrument produces two types of data, Level-1A and Level-1B. Whereas the L1A data are reconstructed, unprocessed instrument data, L1B data are radiometrically and geometrically corrected. Each ASTER data set contains 15 data arrays representing the intensity values from 15 spectral bands (see the ASTER-web page for more detailed information) and various other items of information such as location, date and time. The raw satellite data can be purchased from the USGS online store:

```
https://wist.echo.nasa.gov/wist-bin/api/ims.cgi
```

On this webpage we can select a discipline/topic (e.g., *Land: ASTER*), and then choose from the list of data sets (e.g., *DEM, Level 1A* or *1B* data), define the search area, and click *Start Search*. The system now needs a few minutes to list all relevant data sets. A list of data sets, including various types of additional information (cloud coverage, exposure date, latitude and longitude), can be obtained by clicking on *List Data Granules*. Furthermore, a low resolution preview can be accessed by selecting *Image*. Having purchased a particular data set, the raw image can be downloaded using a temporary FTP-access. As an example, we process an image from an area in Kenya showing Lake Naivasha. The data are stored in two files

```
naivasha_data.hdf
naivasha_data.hdf.met
```

The first file, which has a size of 111 MB, contains the actual raw data, whereas the second file, which has a size of 100 KB, contains the header, together with all sorts of information about the data. We save both files in our working directory. The Image Processing Toolbox contains various tools for importing and processing files stored in the hierarchical data format (HDF). The graphical user interface (GUI) based import tool for importing certain parts of the raw data is

```
hdftool('naivasha_data.hdf')
```

This command opens a GUI that allows us to browse the content of the HDF-file *naivasha_data.hdf*, obtains all information on the contents, and imports certain frequency bands of the satellite image. Alternatively, the command `hdfread` can be used as a quicker way of accessing image data. An image such as that used in the previous section is typically achieved by computing an RGB composite from the *vnir_Band3n*, *2* and *1* in the data file. We first read the data, as follows:

```
clear

I1 = hdfread('naivasha_data.hdf','VNIR_Band3N',...
    'Fields','ImageData');
I2 = hdfread('naivasha_data.hdf','VNIR_Band2',...
    'Fields','ImageData');
I3 = hdfread('naivasha_data.hdf','VNIR_Band1',...
    'Fields','ImageData');
```

These commands generate three 8-bit image arrays each representing the intensity within a certain infrared (IR) frequency band of a 4200×4100 pixel image. The *vnir_Band3n*, *2* and *1* typically contain much information about lithology (including soils), vegetation and water on the Earth's surface. These bands are therefore usually combined into 24-bit RGB images

```
naivasha_rgb = cat(3,I1,I2,I3);
```

As with the previous examples, the $4200 \times 4100 \times 3$ array can now be displayed using

```
imshow(naivasha_rgb);
```

MATLAB scales the images to fit the computer screen. Exporting the processed image from the Figure Window, we only save the image at the monitor's resolution. To obtain an image at a higher resolution (Fig. 7.2), we use the command

```
imwrite(naivasha_rgb,'naivasha_image_vs1_matlab.tif','tif')
```

This command saves the RGB composite as a TIFF-file of about 50 MB in the working directory, which can then be processed with other software such as Adobe Photoshop.

7.5 Georeferencing Satellite Images

The processed ASTER image does not yet have a coordinate system, and the image therefore needs to be tied to a geographical reference frame (*georeferencing*). The raw data can be loaded and transformed into an RGB composite by typing

```
clear
```

Fig. 7.2 RGB composite of a TERRA-ASTER image using the spectral infrared bands *vnir_Band3n*, *2* and *1*. The result is displayed using imshow. Original image courtesy of NASA/GSFC/METI/ERSDAC/JAROS and U.S./Japan ASTER Science Team.

```
I1 = hdfread('naivasha_data.hdf','VNIR_Band3N',...
   'Fields','ImageData');
I2 = hdfread('naivasha_data.hdf','VNIR_Band2',...
   'Fields','ImageData');
I3 = hdfread('naivasha_data.hdf','VNIR_Band1',...
   'Fields','ImageData');

naivasha_rgb = cat(3,I1,I2,I3);
```

The HDF browser

```
hdftool('naivasha_data.hdf')
```

can be used to extract the geodetic coordinates of the four corners of the image. This information is contained in the header of the HDF file. Having launched the HDF tool, we select on the uppermost directory called *naivasha_data.hdf* and find a long list of file attributes in the upper right panel of the GUI, one of which is *productmetadata.0*, which includes the attribute *scenefourcorners* that contains the following information:

```
upperleft  = [-0.319922, 36.214332];
upperright = [-0.400443, 36.770406];
lowerleft  = [-0.878267, 36.096003];
lowerright = [-0.958743, 36.652213];
```

These two-element vectors can be collected into a single array `inputpoints`. Subsequently, the left and right columns can be flipped in order to have *x=longitudes* and *y=latitudes*.

```
inputpoints(1,:) = upperleft;
inputpoints(2,:) = lowerleft;
inputpoints(3,:) = upperright;
inputpoints(4,:) = lowerright;
inputpoints = fliplr(inputpoints);
```

The four corners of the image correspond to the pixels in the four corners of the image, which we store in a variable named `basepoints`.

```
basepoints(1,:) = [1,4200];
basepoints(2,:) = [1,1];
basepoints(3,:) = [4100,4200];
basepoints(4,:) = [4100,1];
```

The function `cp2tform` now takes the pairs of control points, `inputpoints` and `basepoints`, and uses them to infer a spatial transformation matrix `tform`.

```
tform = cp2tform(inputpoints,basepoints,'affine');
```

This transformation can be applied to the original RGB composite `naivasha_rgb` in order to obtain a georeferenced version of the satellite image `newnaivasha_rgb`.

```
[newnaivasha_rgb,x,y] = imtransform(naivasha_rgb,tform);
```

An appropriate grid for the image can now be computed. The grid is typically defined by the minimum and maximum values for the longitude and latitude. The vector increments are then obtained by dividing the longitude and latitude range by the array dimension and by subtracting one from the result. Note the difference between the MATLAB numbering convention and the common coding of maps used in the literature. The north/south suffix is generally replaced by a negative sign for south, whereas MATLAB coding conventions require negative signs for north.

```
X = 36.096003 : (36.770406-36.096003)/8569 : 36.770406;
Y =  0.319922 : ( 0.958743- 0.319922)/8400 :  0.958743;
```

The georeferenced image is displayed with coordinates on the axes and a superimposed grid (Fig. 7.3).

```
imshow(newnaivasha_rgb,'XData',X,'YData',Y), axis on, grid on
xlabel('Longitude'), ylabel('Latitude')
title('Georeferenced ASTER Image')
```

Fig. 7.3 Geoferenced RGB composite of a TERRA-ASTER image using the infrared bands *vnir_Band3n, 2* and *1*. The result is displayed using imshow. Original image courtesy of NASA/GSFC/METI/ERSDAC/JAROS and U.S./Japan ASTER Science Team.

Exporting the image is possible in many different ways, for example using

```
print -djpeg70 -r600 naivasha_georef_vsl_matlab.jpg
```

to export it as a 1 MB JPEG file compressed to 70 %, with a resolution of 600 dpi. Alternatively, we can save it as a TIFF file with a resolution of 600 dpi

```
print -dtiff -r600 naivasha_georef_vsl_matlab.tif
```

which has a size of about 25 MB.

7.6   Digitizing from the Screen: From Pixel to Vector

On-screen digitizing is a widely-used image processing technique. While practical digitizer tablets exist in all formats and sizes, most people prefer digitizing vector data from the screen. Examples for this type of application include digitizing river networks and catchment areas on topographic maps, the outlines of lithologic units on geological maps, the distribution of landslides on satellite images, and the distribution of mineral grains in a microscope image. The digitizing procedure consists of the following steps. Firstly, the image is imported into the workspace. A coordinate system is then defined, allowing the objects of interest to be entered by moving a cursor or cross hair and clicking the mouse button. The result is a two-dimensional array of xy data, such as longitudes and latitudes of the corner points of a polygon or the coordinates of the objects of interest in a particular area.

The function `ginput` included in the standard MATLAB toolbox allows graphical input using a mouse on the screen. It is generally used to select points, such as specific data points, from a figure created by an arbitrary graphics function such as `plot`. The function `ginput` is often used for interactive plotting, i.e., the digitized points appear on the screen after they have been selected. The disadvantage of the function is that it does not provide coordinate referencing on an image. We therefore use a modified version of the function, which allows an image to be referenced to an arbitrary rectangular coordinate system. Save the following code for this modified version of the function `ginput` in a text file *minput.m*.

```
function data = minput(imagefile)
% Specify the limits of the image
xmin = input('Specify xmin! ');
xmax = input('Specify xmax! ');
ymin = input('Specify ymin! ');
ymax = input('Specify ymax! ');
```

```
% Read image and display
B = imread(imagefile);
a = size(B,2); b = size(B,1);
imshow(B);

% Define lower left and upper right corner of image
disp('Click on lower left and upper right corner, then <return>')
[xcr,ycr] = ginput;
XMIN = xmin-((xmax-xmin)*xcr(1,1)/(xcr(2,1)-xcr(1,1)));
XMAX = xmax+((xmax-xmin)*(a-xcr(2,1))/(xcr(2,1)-xcr(1,1)));
YMIN = ymin-((ymax-ymin)*ycr(1,1)/(ycr(2,1)-ycr(1,1)));
YMAX = ymax+((ymax-ymin)*(b-ycr(2,1))/(ycr(2,1)-ycr(1,1)));

% Digitize data points
disp('Click on data points to digitize, then <return>')
[xdata,ydata] = ginput;
XDATA = XMIN + ((XMAX-XMIN)*xdata/size(B,2));
YDATA = YMIN + ((YMAX-YMIN)*ydata/size(B,1));
data(:,1) = XDATA; data(:,2) = YDATA;
```

The function `minput` has four stages. In the first stage, the user enters the limits of the coordinate axes as reference points for the image. Next, the image is imported into the workspace and displayed on the screen. The third stage uses `ginput` to define the upper left and lower right corners of the image. In the fourth stage the relationship between the coordinates of the two corners on the figure window and the reference coordinate system is then used to compute the transformation for all of the digitized points.

As an example, we use the image stored in the file *naivasha_georef. jpg* and digitize the outline of Lake Naivasha in the center of the image. We activate the new function `minput` from the Command Window using the commands

```
clear

data = minput('naivasha_georef_vs1_matlab.jpg')
```

The function first asks for the coordinates for the limits of the *x*- and *y*-axis for the reference frame. We enter the corresponding numbers and press *return* after each input.

```
Specify xmin! 36.1
Specify xmax! 36.7
Specify ymin! -1
Specify ymax! -0.3
```

The function then reads the file *naivasha_georef_vs1_matlab.jpg* and displays the image. We ignore the warning

```
Warning: Image is too big to fit on screen; displaying at 33%
```

and wait for the next response

```
Click on lower left and upper right corner, then <return>
```

The image window can be scaled according to user preference. Clicking on the lower left and upper right corners defines the dimension of the image. These changes are registered by pressing *return*. The routine then references the image to the coordinate system and waits for the input of the points we wish to digitize from the image.

```
Click on data points to digitize, then <return>
```

We finish the input by again pressing *return*. The *xy* coordinates of our digitized points are now stored in the variable data. We can now use these vector data for other applications.

Recommended Reading

Abrams M, Hook S (2002) ASTER User Handbook - Version 2. Jet Propulsion Laboratory and EROS Data Center, Sioux Falls

Campbell JB (2002) Introduction to Remote Sensing. Taylor & Francis, London

Gonzalez RC, Woods RE, Eddins SL (2009) Digital Image Processing Using MATLAB – 2nd Edition. Gatesmark Publishing, LLC

The Mathworks (2010) Image Processing Toolbox – User's Guide. The MathWorks, Natick, MA

8 Editing Graphics, Text, and Tables

8.1 Introduction

In Chapters 2 and 3 we extracted text and tables from journal articles, web-pages, and online data bases. In Chapters 5 and 6 we then created various simple line graphs, bar plots, and block diagrams with MATLAB. In Chapter 7 we processed images, in particular satellite imagery. Both the vector graphics created in Chapter 5 and the raster graphics generated in Chapters 6 and 7 require further editing before they can be published in journal articles or books, presented as elements of posters, or included in conference presentations. In this chapter we demonstrate how to edit vector graphics, raster graphics, text, and tables using both open source and commercial software packages. Section 8.2 is on editing and improving simple line graphs, as an example of vector graphics. The properties of the MATLAB figures, such as their line and fill colors, background color, stroke width and appearance, font type and size, and other properties, are modified to improve the legibility of the graphics. Section 8.3 demonstrates how to improve raster images by changing colors, color mode, resolution and other properties, and how to export the results in a specific file format that will depend on the type of document in which the image is to be used. Section 8.4 is about formatting text, structuring and tagging manuscripts, consistent use of paragraph and character styles, fonts, non-breaking spaces and other special characters. Section 8.5 is about editing tabulated data to prepare it for further formatting.

The following section provides a very brief introduction to the use of typical graphics editors by means of step-by-step tutorials, supplemented by screenshots documenting the workflow that are provided on the CD accompanying this book. In this chapter all files with filenames ending in ..._matlab.eps were created with MATLAB in Chapters 5 or 6, and are therefore stored in the materials folders of these chapters on the CD. The software used in this chapter was introduced in Chapter 1 and includes state-of-the-art graphics tools such as *Adobe Illustrator* and *Photoshop*,

M. H. Trauth and E. Sillmann, *MATLAB® and Design Recipes for Earth Sciences*,
DOI: 10.1007/978-3-642-32544-1_8, © Springer-Verlag Berlin Heidelberg 2013

as well as widely used and very popular *Microsoft Word* and *Apple Pages*, and freely available open-source software such as *OpenOffice, Inkscape* and *Gimp*. After learning how to edit vector and raster graphics, texts, and tables we will then be ready to move on to the last three chapters (Chapters 9 to 11), which are on presentations, posters, and manuscripts.

8.2   Editing Vector Graphics

For our first example in this chapter we will demonstrate how to edit vector graphics. Vector graphics use geometrical primitives such as polygons, patches (or filled polygons), lines, or points. Examples of vector graphics are the line graphs, bar plots, and pie charts created in Chapter 5, and the coastline plot from Chapter 6. In MATLAB, simple lists of *xy* coordinates for *NaN*-separated polygons (such as the *age-temperature* data from Section 5.2, or the GSHHS Shoreline Data Set introduced in Section 6.2, form the basis of vector graphics. When using MATLAB or any other software to visualize such data, the graphics functions display plots of the *xy* coordinates as complex vector graphs with multiple graphics properties that include as line thicknesses and colors, markers, labels, background colors, coordinate axes, and so forth. Vector graphics software then allows the properties to be selected and other, more sophisticated properties, to be added to be graphics.

Line Graphs: Getting Started, Palettes, Layers, and Tools

In this subsection we will first edit the line graph from Section 5.2 showing temperature and snow accumulation variations in the GISP2 data from R.B. Alley (2000). Comparing the original line plot created with MATLAB with the totally redesigned versions shown in Figure 8.1 gives an idea of the possibilities described in this section.

We use both the free *Inkscape* software and the commercial *Adobe Illustrator* to demonstrate some of the basics of editing vector graphics. All vector graphics in this book have been created with the English editions of Inkscape (v0.48) or Adobe Creative Suite CS5. Unfortunately, CS5 files can not be opened and edited with versions older than CS4. Our tutorial, however, should also work in a very similar way with older versions of the software, or with versions in other languages. Further information about the software and technical terms or jargon in italics can be found at

http://help.adobe.com/en_US/illustrator/cs/using/index.html

and

```
http://wiki.inkscape.org/wiki/index.php/Help:Contents
```

Both the Inkscape and Illustrator programs allow us to import the EPS file *icecore_lineplot_vs1_matlab.eps* created with MATLAB in Chapter 5 using *Open File* from the *File* menu. Most vector graphics software tools should not have a problem with opening an EPS file; in the rare case of incompatibility problems, we could also create PS or PDF files with MATLAB and import these, again using *Open File*.

Both programs display the line graph in a new figure window. At the top of this window we find the *Program Menu*. In this drop-down menu we find a number of push-buttons assembled in a toolbar. These provide access to all kinds of information about the document and also include various shortcut buttons for some frequently-used functions, and various context commands. On the left side of the window we find the *Tools* palette. To become familiar with the names and functions in the toolbar we can move the mouse over their buttons to obtain some information on the functions.

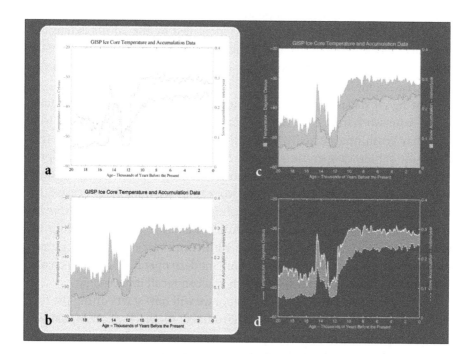

EDITING GRAPHICS, TEXT, AND TABLES

8

Fig. 8.1 Original and Variations of the GISP2 data from R.B. Alley (2000). **a** original MATLAB plot, **b** filled graph, **c** negative, for dark background, **d** negative, for a presentation or leaflet

In contrast to Illustrator, Inkscape also assists in the use of a graphics tool by providing a short description of the tool while it is being used.

Having opened the *icecore_lineplot_vs1_matlab.eps* file, we immediately save the document under a new name using *Save As* from the *File* menu. We call our Inkscape file *icecore_lineplot_vs2_inkscape.eps* and our Illustrator file *icecore_lineplot_vs3_ai.eps*. It is recommended that the original MATLAB file should always be retained, as we may wish to return to it at a later stage to create another version of the graph. Both Inkscape and Illustrator allow us to save the file in various vector file formats, such as EPS, PS or PDF, and these are the formats that we will use because they are compatible with most other vector graphics programs, whereas vector file types that are native to particular types of software can often not be imported into other programs. The default file type in Inkscape is the *Scalable Vector Graphics* (SVG) format, which is an XML-based file format that can be edited with a text editor. The *XML Editor* from the *File* menu in Inkscape provides a quick view of the source code of the XML file. We will not, however, use this source-code editing feature of Inkscape in our example, but will instead use the GUI-based WYSIWYG (*what you see is what you get*) graphics editor to manipulate our graphics documents. Illustrator has its own native file format, which is identified by the *.ai* file extension. Illustrator files are compatible with other vector graphics tools when you include the PDF information of the graph in the file; this is an optional feature when creating files in Illustrator.

During all stages of our graphics design project it is always important to save your work from time to time using *Save as*, since any graphics software may crash unexpectedly, especially when using older versions of the Microsoft Windows operating system. More recent operating systems provide an automatic save every few minutes, making it unnecessary to save your files manually. Backing up your work onto separate disks every 5 to 10 minutes is however still advisable, using, for example, *Time Machine* in Mac OS X, or another automated backup system. It is also a good idea to always keep earlier versions of a graph while working on it, since you may wish to return to a previous layout. Again, modern operating systems create different versions of all your files, which you can recall if necessary.

Having imported and saved the graph, we will now start editing the file. We first need to set up our working environment (or workspace) so that we can quickly access the most frequently used functions. In Illustrator, we select *Like Photoshop* from the *Window > Workspace* menu to open all the panels that we wish to work with (Fig. 8.2). In Inkscape (Fig. 8.3), we open the *Layers* dialog window using *Menu Bar > Layer > Layers*, or alternatively,

Fig. 8.2 Screenshot Workspace Adobe Illustrator, showing menu, drop down menue *Window*, toolbar, navigator, *Swatches* panel, *Appearance* panel, *Layers* panel and the document window.

using the shortcut button *Commands Bar > Toolbar > View Layers*. Still in Inkscape, we also launch the *Undo History Dialog* and the *Fill and Stroke Dialog* boxes which we will need when editing the line graph. Taking a closer look at the *Layers* panel in Illustrator or *Layers* dialog in Inkscape, we note that the three original graphics layers created by MATLAB, *figure window*, *axes*, and *plot objects*, were unfortunately not preserved as separate layers when the graph was saved as an EPS file. In both programs, all of the graphics objects were instead merged into one single layer. For further editing of the graphics we need to move all different types of object into individual layers. The advantage of working with layers is that the different types of objects can be selected and edited individually by toggling the lock and visibility symbols of each layer. Illustrator also displays a list of all the objects within a single layer in the *Layers* panel when you click on the grey triangle next to the layer's name, e.g., next to *Layer 1*. As you can see, there are many different types of graphics elements in *Layer 1*, such as *<Text>*,

Fig. 8.3 Screenshot of the Inkscape Workspace, showing *Menu, Commands* bar with general command buttons, *Tool Controls* bar, *Undo History Dialog, Colors* bar, and the *Layers* panel. The *Status* bar at the bottom of the window displays useful hints during work.

<Path> and *<Group>*, which can be selected individually by clicking the corresponding element in the list, e.g. *<Path>*, which then becomes active and is marked with a small square.

Editing a graph in both programs generally includes the following steps: (1) Creating and naming layers for different graphics objects, such as the text, the line for snow accumulation, the line for temperature, the coordinate system (with ticks and axis labels), the background, and the frame; (2) selecting and dragging the individual objects into the specific layers; (3) editing the white background and frame; (4) changing the properties of the graphics objects, such as their line styles, fillings, and colors; (5) editing text objects in the layers, e.g., choosing serif or sans-serif fonts and font sizes; (6) adjusting technical properties affecting the printing process, such as the handling of transparency and the color black; and finally (7) saving the resulting graph as a PDF or an EPS file.

We start editing the line graph by creating a new layer in *Inkscape* using

Layer > Add Layer, naming the new layer *text*, selecting all text elements using the uppermost tool in the *Select and Transform Objects* vertical toolbar, and moving all text objects to the new layer using *Layer > Move Selection to Layer Above*. We next create a second new layer named *graph* in a similar manner, select the two curves, and move them into this second layer. To check that text elements and curves indeed arrived in the designated layers, we can click the eye symbol to toggle the visibility of the layers. The different elements of the coordinate axes can not be selected individually as they are all grouped together. Grouping graphics elements is a very useful feature of most vector graphics software tools that allows groups of objects to be moved without changing the arrangement of the individual objects within the group. We can ungroup the different elements of the coordinate axes by using *Object > Ungroup*. As an example of further editing of the coordinate axes, we find a superfluous blue tick located above the black line in the upper right corner of the coordinate axes that we can remove by first ungrouping all elements, and then removing the object using the *backspace* key. We then repeat this procedure for the other three corners of the coordinate axes, to remove other superfluous ticks.

After moving the blue axis to the layer named *coordinate* and toggling the visibility of this layer we come across another interesting detail. MATLAB has obviously generated two lines in the position of the blue *Snow Accumulation* axis, i.e. there is a second, hidden, red axis lying behind the blue axis. We continue to activate and move each type of object into separate layers until we have created the layers *text*, *graph accumulation*, *graph temperature*, *coordinate system & ticks*, *back*, and *back2/frame*. To deselect all selected objects we simply click with the pointer on an empty space in the work area. Whenever we wish to undo one of our steps, we can always do so using the *Undo History* dialog box.

Organizing objects into layers works in a very similar way with *Illustrator*, with just a few small differences. The *Layers* panel of this software allows graphics objects belonging to individual layers to be organized and viewed in series of sublayers. These sublayers can be viewed by clicking with the pointer on the small triangle to the left of the particular layer of interest. A new layer is created by clicking on the upper right corner of the *Layers* panel to open the *Layers* menu, and then selecting *New Layer*. A small dialog window opens for the new layer in which we can choose a color for the layer, which allows objects belonging to this particular layer to be easily identified. In Illustrator, a polygon or vector is called a *Path*. The *Selection Tool* is found in the vertical toolbar, represented by a black arrow. Clicking on the line of the graph makes the *curve anchors* visible and shows a small square on the right in the *Layers* panel, filled with the same color as the curve to show it is active. We find that the curves created by MATLAB have unfortunately each been split into three segments, which we need to select individually and drag into the new layer. We do this by selecting the line segments, selecting the colored square in the panel, and moving both into the designated layer, which we then protect by toggling the lock symbol.

Having moved the various elements of the axes and curves into separate layers, we do the same with the text. We can select text very quickly using *Menu Bar > Select > Object > Text Objects* and moving the colored square that appears in the *Layers* panel up to the *text* layer. Having first ensured that the *graph* layer is locked, we choose *Toolbar > Direct Selection Tool* and very quickly select the black, and the blue or red, axes and ticks by clicking on one black tick and then selecting *Menu Bar > Select > Same > Stroke color*. By toggling the visibility symbol in the *Layers* panel we can check in which layer a particular object is located. To move one layer to the background or foreground, we click and drag the whole panel row of this layer to a higher or lower position in the *Layers* panel. If we want to undo anything that we have done, we can step backwards using *Menu Bar > Edit > Undo* (and forwards again using *Redo*).

MATLAB seems to take account the fact that the *PostScript* standard only supports certain specific techniques for implementing transparency. In both Illustrator and Inkscape, the EPS document contains two white rectangles. We move the larger of these two rectangles to a layer named *frame* which we then move down to the bottom of the list of layers. The smaller rectangle has the same dimensions as the coordinate axes. We move it to the *back* layer, which we then move to the second lowest position in the list of layers. We can also edit the appearance of the two rectangles in *back* and *frame* by, for instance, changing the *Fill* color or the *Stroke* color, or modifying the sizes of the rectangles. After moving the two rectangles into the *back* and *frame layers*, we can lock these layers and then hide them by switching off their visibility. We save the line drawing once again as *icecore_lineplot_vs2_inkscape.eps* in Inkscape or, if you work with Illustrator, as *icecore_lineplot_vs3_ai.eps*.

We next continue with editing the curves themselves by, for example, filling the area between the curves and between the *x*-axis and the graph (Fig. 8.1 b and c). Before we are able to fill in this area we need to join together the three segments of the curves. This splitting into separate segments is an unfortunate characteristic of polygons created by the PostScript export plugins of many types of software, including MATLAB. Merging the three polygon segments is performed in Illustrator by selecting the three segments of the blue line and then combining them using *Menu Bar > Object > Compound Path*. Alternatively, to merge two segments of the blue polygon, we can use the *Direct Selection* tool and select one of the segments. Using the *Lasso* tool, we carefully select the terminal vertex at the end of the segment of the blue line and the vertex at the beginning of the adjacent segment. Using *Menu Bar > Path > Join* we combine both segments into a single segment, and then repeat the process with the remaining third segment. The segments of the red curve are then linked using the same procedure.

As the final step in editing the line graph, we will now fill in the area beneath the curves. For this, we duplicate the *graph* layer using *Layer Panel Menu > Duplicate*, rename the new layer *graph fill*, put the layer into the third-lowest position in the list of layers, unlock the new layer, and lock all other layers. To create a closed polygon, we now give a demonstration of how to use Illustrator to create new polygons, which of course is the classic application of vector graphics software tools. We choose *Pen Tool* from the *Toolbar* and select the left terminal vertex of the blue curve, which is indicated by / *anchor* when hovering over this vertex. Holding down the *Shift* key on the keyboard allows the creation of a new vertex located either vertically in 90° angle above or below, or horizontally in 180° angle to the left or right, of the terminal vertex. We decide to place the new vertex at the lower end of the red axis, on the left hand side. While still holding the *Shift* button down, we click on the right end of the black *x*-axis and a third time at the same height as the right end of the blue curve, trying to meet its terminal vertex. Taking a closer look at the snow accumulation curve, we note that there is no data point for zero age, i.e. a small gap exists between the blue curve and the blue *y*-axis to the right, which means that the vertical line does not meet the right end of the blue curve. To close this gap, we connect the blue curve and our new vertical line using the *Lasso* tool as described above.

Next, to select a color shade for the *Fill* that goes together nicely with the blue of the curve, we use the *Swatches* panel and the *Color Guide* panel. The *Swatches* panel should already be visible in our workspace. We can browse all shades of colors using *Menu Bar > Window > Color Guide*. We again select the blue curve, and then the *Tools* panel shows which color is used for the *Stroke* of our object. By clicking on the small arrow in the right corner (next to the symbols for stroke and fill in the *Tools* panel) we switch between stroke and fill to apply an intense blue color to the fill. As the color is a bit too intense, we select a paler shade from the *Color Guide* panel, and the fill, still being active, changes color.

We then transfer the colors in our document to the *Swatches* panel. To

do this we unlock all layers using the *Layers* panel menu and *Menu Bar >
Select > All*, after which the colored symbols for each layer appear as small
squares. We click on the upper right corner of the *Swatches* panel, open its
Context menu, and choose *Add Selected Colors,* which automatically adds
all colors used in the document, each represented by a square color field.
We continue by creating the color fill for the area beneath the red curve and
choosing a suitable shade, as we did for the blue curve. We save the file as
icecore_lineplot_vs5_ai_fill.eps.

If the graph is to be published in a research article, the printing costs
can be reduced by using a black and white (b/w) version. Printed versions of
papers today often contain b/w illustrations, with colored versions included
in the PDF version of the paper published online. It is important to note that
a figure in *RGB* color mode, which is used in digital versions to be read on
a computer screen, need to be converted into the *CMYK* color mode when
the figure is being provided for offset printing. In other words, the black
color should be a pure black (also known as *key*), and not a composite of
maximum intensity red, green, and blue in the RGB mode. We can change
the color mode into CMYK using *Menu Bar > Edit > Color Mode* and then
check the two black color swatch buttons in the *Swatches* panel. Double
clicking on a color swatch opens a panel called *Swatches Options*. We find
that the ticks and x-axes use the correct color K=100 (for 100 %), and remove
the other black color by selecting it and then clicking on the trash icon in
the lower right corner of the panel. We then select any of the black text objects and automatically select all others using *Menu Bar > Select > Same >
Fill color*, and then use K=100 for all text in the graph. If we do not change
RGB black into CMYK black (or *key*), all black text will appear as a smeared
mix of the three colors, red, green and blue, since the ability to print the
three colors precisely on top of each other is very rare in professional printing. We adjust the color values for red and blue to whole numbers and save
the file as *icecore_lineplot_vs6_ai_cmyk.eps*.

Continuing with the grey scale version of the graph *icecore_lineplot_
vs5_ai_fill.eps*, we select all graphics objects, such as coordinate axes and
ticks, and change their stroke color into CMYK black (or *key*). Double clicking on the color swatch in the *Swatches* panel opens a small window called
Swatches Options. Instead of *Color Type > Process Color*, we now choose
Grayscale instead. To deactivate all objects after we have completed this task,
we click with the pointer on an empty space in the work area. Since there is
now no color information available to distinguish between the two curves,
which were originally red and blue, we select the *Snow Accumulation* curve
and change the line to a dashed line using the *Contour Style* dialog box

of the *Contour* panel. Alternatively, we could again use different shades of grey for the filled polygons, i.e. the areas below the curves, to distinguish between the two curves. In order to associate each of the two curves with its corresponding *y*-axis, we can either label the curves or add a legend. We then save the file as *icecore_lineplot_vs7_ai_greyscale.eps*.

We have demonstrated a typical workflow for editing a line graph, which can also be used to edit other *xy* and *xyy* plots. As an example, the bar plot created in Section 5.3 stored in *icecore_bargraph_vs1_matlab.eps* essentially comprises similar graphics objects, such as the axes, text objects, and the white background, but with filled rectangles with black outlines instead of solid lines.

Pie Charts: Transparency, Raster with Vector, and Advanced Features

The pie chart from Chapter 5 in *icecore_piechart_vs1_matlab.eps*, visualizing 2D data in 3D, looks very attractive when using a transparency effect. However, if we have already generated transparency in MATLAB the program unfortunately exports a raster file (even though it was saved as the EPS file *icecore_piechart_vs2_matlab.eps*) and the vector information is lost. Of course for many applications a raster (or bitmap) file is easier to handle than a vector file in a production process. However, disadvantages of the bitmap format include the greater file size, the fact that no further editing of text or other objects is possible, the fixed resolution, and the very limited cropping and editing options of a raster file. We might therefore prefer to export an EPS file from MATLAB with opaque colors, such as in *icecore_piechart_vs1_matlab.eps*, and then add transparency with the vector graphics software tool (Fig. 8.4). To demonstrate this, we open the file in Illustrator, choose *Toolbar > Direct Selection Tool*, select all objects using *Menu Bar > Select > All*, and immediately reveal the complex structure of this chart. With all objects still selected we click on the *Swatches* panel, open its context menu, and choose *Add Used Colors*. To deactivate all selections we click with the pointer over empty space of the work area.

Using *Toolbar > Direct Selection Tool*, we click and drag the text object *>1 %* a few millimeters to the right to avoid overlapping text. If we aim to publish the graphic in a research article, we can change the color mode for the whole document into CMYK using *Menu Bar > Document Color Mode > CMYK*. We then select all text objects and move them into a separate layer called *text*, using *Menu Bar > Object > Text Objects*. With all text objects still selected, we change the color mode in the *Color* panel to CMYK and find that the black color for text objects is still RGB black, consisting of a

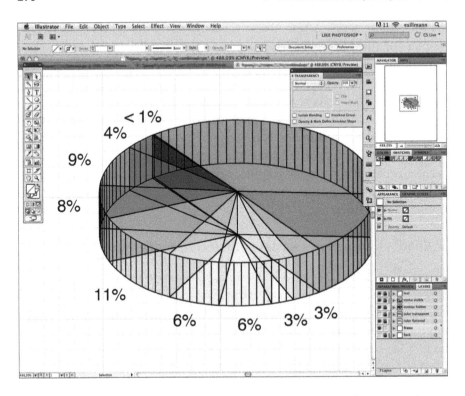

Fig. 8.4 Screenshot of the icecore pie chart figure, showing the Adobe Illustrator workspace with *Menu* bar, *Toolbar* and among others, the panels *Navigator, Swatches, Appearance* and *Layers*.

mix of the three colors red, green and blue, as in the previous graph. We therefore again have to change the color composition by moving the scrollbar of each color channel to the values C = 0, M = 0, Y = 0, K = 100, or alternatively, by typing the values for CMYK manually into the corresponding text fields. Clicking on the color field in the *Swatches* panel opens a small window entitled *Swatch Options*. Instead of *Color Type > Process Color*, we now choose *Grayscale*, as we did in the previous example. To deactivate all objects we click the pointer over an empty space of the work area. We then select all black outlines using the *Direct Selection Tool*, and then using *Menu Bar > Select > Same > Stroke color*, after which we move them into a new layer called *contours*. The black outline color is again RGB black and we therefore have to delete the mixed RGB color in the *Swatches* panel. All contours are changed into the correct grayscale black by clicking on the correct black color field.

We can now examine the results of our work by toggling the visibility and lock for the new *text* and *contours* layers. Using *Menu Bar > Select > All*, the colored zones and two rectangles in the background are selected, as in the line graph discussed previously. We again drag the rectangles into a new layer called *back*, move this layer to the lowermost position in the *Layers* panel, and lock the layer. Using *Menu Bar > Select > All* should now only select the colored zones, and we then move them into a new layer called *color*. The original layer is now empty and can be deleted. Clicking on the *color* layer, we again activate all colored objects in our pie chart and can change the opacity using the *Transparency* panel, either by typing *50 % Opacity* or by dragging the scrollbar to the appropriate value. We then save the file using *Menu Bar > File > Save As*, choose *Medium Resolution* or *High Resolution* depending on our needs, activate the *Compatible Gradient and Gradient Mesh Printing* checkbox, and then save the file under the name of *icecore_piechart_vs4_ai_transparency.eps*. Having completed all of these steps the pie chart is now ready for digital printing, or for inclusion in any type of presentation.

Let us now continue with some of the advanced features offered by Illustrator for creating files suitable for offset printing. In theory, creating files by flattening gradients such as transparency is always possible by exporting the whole graphic as a bitmap, or in raster form with the required resolution, using *Menu Bar > File > Export*, for example as the JPEG file *icecore_piechart_vs5_ai.jpg*. A second possible way to create a vector file (which is not recommended in every case) is to transform each area of different color into an individual object using *Menu Bar > Object > Flatten Transparency*. The third and safe method to use for offset printing is to first export the critical transparent areas into a bitmap file, and then later replace them in the illustration, thus retaining the text and contours as vectors. We continue with the *icecore_piechart_vs4_ai_transparency.eps* file that we created previously, and toggle the visibility so that only the *color* layer remains visible. Using *Menu Bar > File > Export ...* we export the file as, for instance, a Portable Network Graphics (PNG) file with a high resolution. We then toggle the visibility again so that only the *color* layer remains hidden. This layer will be eliminated at a later stage. We then form a new layer called *color bitmap* and import our PNG file into it using *Menu Bar > File > Place ...*, move the image to the correct position, so that it replaces the image in the *color* layer, and save the file as *icecore_piechart_vs10_ai_combined. eps*.

As a second example of an advanced application, let us create a cropped image from our pie chart using the *clipping path* feature in Illustrator. We first need to define the outline of the cropped image using the *Rectangle tool* from the *Toolbar*. With this tool we draw a rectangular outline around the diagram, clicking on one of its corners, then drag the cursor diagonally until the rectangle has the same width as the pie chart, but a slightly lower height. Having completed the rectangle, we use the *Add Anchor* tool from the *Toolbar* to add an anchor point to the middle of each of the two horizontal sides of the rectangle, and then drag these anchor points to the upper and lower edges of the pie chart. To convert this irregular-shaped hexagon into the pie shape we then use the *Convert Anchor Point* tool from the *Toolbar*, click on the upper point, and drag the handle of the *Bezier* curve to transform the horizontal line into a semicircle. After also adjusting the lower horizontal line, we select this irregular-shaped hexagon and the PNG image of the pie. Using *Menu Bar > Object > Clipping Mask > Make* then creates a *clipping path* following the outline of the PNG image, which cuts away the white background that surrounds the original image. To see whether the *clipping path* does in fact accurately cut away the white parts, we can draw a random shape with an arbitrary fill in the background and delete it after this check. Finally, we save our file in an EPS format as *icecore_piechart_vs11_ai_mask.eps*.

Our third advanced example demonstrates *overprinting* of black contours and text objects during the offset printing process. Using *Menu Bar > Window > Separations Preview*, we open a panel displaying the four printing plates for *Cyan, Magenta, Yellow and Key/Black* (CMYK). If we hide the black printing plate, we notice that, in little gaps between the colored areas, lines and text, that were previously black, now appear white in the *Separations Preview*. Since the four printing plates can become slightly displaced relative to each other during printing, the final product can have black lines or text with white shadows or ghosts. The avoid this problem, we select all the black objects in our illustration by first selecting any arbitrary black object and then select all others using *Menu Bar > Select > Same > Stroke Color*. Using *Menu Bar > Window > Attributes* we then open a new panel, in which we click on the *Overprint Stroke* checkbox. The result can be seen by again hiding the black printing plate in the *Separations Preview* panel: the colored areas are now continuous beneath the black contours. Similarly, we add the overprinting feature to the black text objects by activating *Overprint Fill* checkbox in the *Attributes* panel, and save the final product as a file with the name *icecore_piechart_vs12_ai_overprint.eps*. Please note that the overprinting attribute needs to be handled with care. In most cases, it is only applied to black objects since overprinting for lighter colors, in particular for white, leads to the complete invisibility of objects in these colors.

As our last example of the advanced features in Illustrator, we save the colors used in this particular file for use in future applications, using the *Adobe Swatch Exchange* (ASE) format. We first open the *Color* panel menu by clicking on the triangle in the upper right corner of the *Color* panel. Choosing *Add Used Colors* transfers all swatches into our panel. Opening the *Color* panel menu again, we choose *Small List View* and now find all colors listed by their names, for instance, $C = 0$, $M = 0$, $Y = 0$, $K = 100$ for the black color, named after the color separation. Clicking on the black color in the *Color* panel opens the *Swatch Options* window to allow various options to be checked or adjusted, e.g., once again choose grayscale for the $K = 100$ color. Back to the *Color* panel menu, we click on *Save Swatch Library as ASE* to save the colors as *icecore_piechart_swatches.ase* so that this *Swatch Library* can be transferred to other graphics files. If we want to load a *Swatch Library* into another graphics file, we select *Open Swatch Library* in the menu and search for the desired file on our computer or use the built-in library. The swatch library saved in *icecore_piechart_swatches.ase* is now available for use with other applications within the entire Adobe Creative Suite, which includes not only Illustrator but also InDesign and Photoshop, or can be handed over to another colleague for further processing.

Rose Diagrams: Manipulating Clipping Paths

The rose diagram in the *directional_rosediagram_vs1_matlab.eps* file reveals that MATLAB has created a rather complex assemblage of graphics objects such as lines, patches, and background rectangles, which will have to be edited before they can be included in a publication or presentation. Clicking on the triangle symbols on the left side of the *Layer* panel shows one group that includes the red polygon, the text elements, and four other groups containing various graphics objects, various paths, and the clipping paths. We select and organize the text and the red polygon into separate new layers called *text* and *graph*, as we did in the previous sections.

Zooming into the upper and lower parts of the diagram, we find that the outer circle extends beyond one of the object's frames, or more precisely, beyond the limits of one of the clipping paths. By toggling the visibility of each of the layers one by one, we check in which layers the different objects are located and find that hiding all clipping paths will not change the overall appearance, but at least fixes the truncated outer circle. We create new layers called *dotted radiation*, *outline*, and *dotted circles* and move the relevant objects into these layers. All superfluous objects and clipping paths of the graphic remain in the original *Layer 1*, at the lowest level in the *Layer* panel. We toggle the *Layer 1* to make it invisible; later it will even be deleted.

Clicking on the *graph* layer activates the red polygon and we choose a white color for the fill. Opening the *Stroke* panel and clicking on the content of the *dotted circle* and *dotted radiation* layers, we find that a *stroke weight* of 1 pt (points) has been used, in combination with a *dashed line* that has a 0.5 pt dash and 4 pt gap, both of which settings we can change to other values in this panel.

We then check the appearance of the diagram, which overlies a colored background, by creating a new layer at the bottom of the list of layers and drawing a filled rectangle using *Toolbar > Rectangle Tool*. After this we again delete this layer. We might prefer a white background for the rose diagram circle: we therefore duplicate the *outline* layer and name it *white circle*. We then lock all layers except the new one, move it below the original *outline* layer, set the *white circle* contour to *None (/)*, and give the filling a white color. To complete our procedure we remove the *back* and *Layer 1* layers, and save our file under the name *directional_rosediagram_vs1_ai_lines.eps*. For other purposes we can give the rose diagram circle a fill similar to that in the file named *directional_rosediagram_vs2_ai_fill1.eps* in the CD accompanying this book, or define different colors of each segment of the rose diagram, as in the file called *directional_rosediagram_vs3_ai_fill2.eps*.

Shoreline Data Set: Number of Vertices in Vector Graphics

For our next example we will work on the GSHHS shoreline plot created in Section 6.2. We have already discussed the problem of large numbers of vertices (in Section 6.2), and used a MATLAB script to halve the number of data points plotted. At first sight, the graphics object in the *coastline_linegraph_vs1_matlab_17703vertices.eps* file looks very familiar after having worked on the line graph in *icecore_lineplot_vs1_matlab.eps*. The difference is that the shoreline data set has 17,703 pairs of longitudes and latitudes, resulting in polygons with a total of 17,703 vertices in the graph. Despite the relatively small file size of 56 KB, this enormous number of vertices may cause problems when trying to print the polygons on paper with a PostScript printer, and also when including the graph in slides or another media. For example, a PDF document containing such complex vector objects is difficult to browse through or print, and also difficult to handle as a source for offset printing. As an example we will consider Figure 2 from a publication in Quaternary Science Reviews (Trauth et al. 2009),

Trauth, M.H., Larrasoaña, J.C., Mudelsee, M. (2009):

Trends, rhythms and events in Plio-Pleistocene African
climate. Quaternary Science Reviews, 28, 399-411.

which can be downloaded from the Science Direct webpage, if you have
electronic access to that journal:

http://sciencedirect.com/science/article/pii/S0277379108003302

There are three panels in this figure: A, B and C. Panels B and C comprise
three curves for different paired window sizes, 15 kyr, 30 kyr and 50 kyr.
For mathematical reasons each of these curves consists of 5,000 data points,
i.e. the polygons have 5,000 vertices. Having downloaded the file, the very
tedious scrolling forward and backward across Figure 2 clearly demonstrates the problems of dealing with large numbers of vertices. Furthermore,
printing the file is a great challenge, as well as being a time-consuming task
on most printers. In this example, the copy editing office of the publisher
should have come back to the author to improve the graph for publication.
In order to get around this problem a second version of the file has been
created, which includes a JPEG version of Figure 2 for easier scrolling and
printing. Another example is Figure 5.20 in *MATLAB Recipes for Earth
Sciences–3rd Edition* (Trauth 2010), which was originally delivered to the
designer as vector file with 5,000 vertices for each curve but subsequently
converted into a TIFF file for publication.

To prepare a job for printing, most printing devices use a built-in *Raster
Image Processor* or *RIP* software module to interpret, render, and screen
vector or bitmap data, which essentially means to transform it into a high-
resolution raster image. This process is called *ripping* or *rendering* vector
data. If the processing of a complex vector file approaches or exceeds the
limitations of the device, the printer may require an extremely long processing time, or even stop printing and return a *limit check* or *VMerror* (which
stands for *Virtual Memory Error*) message. Editing such a complex vector
file, or placing it in a desktop publishing document, or embedding it into
PDF may also, in some cases, result in a serious loss of performance, depending on the available hardware and the quality of the software product
used.

In most cases (e.g., when using the graph in a conference presentation,
on a poster, or in a manuscript), processing of the original, high-resolution
shoreline data set is not advisable. If you are the creator of the high-resolution figure in *coastline_linegraph_vs1_matlab_17703vertices.eps*, it
would be better to run the script from Section 6.2 to reduce the number
of vertices. For the sake of comparison, we first run the script once and

reduce the number of vertices by a factor of two, creating a plot with 8,852 vertices, stored in *coastline_linegraph_vs2_matlab_8852vertices.eps*. A second run again halves the number of vertices to just 4,426, and we store this plot in a file called *coastline_linegraph_vs3_matlab_4426vertices.eps*. The original illustration can also be exported from MATLAB as a bitmap file of 600 dpi or, in our example, 300 dpi resolution, creating a file called *coastline_linegraph_vs1_matlab_17703vertices.tiff*.

Should we not have access to the original data set and therefore not be able to run the MATLAB script to reduce the number of vertices, we can instead use graphics software to reduce the number of vertices (referred to as *nodes* and anchor points in graphics software jargon). The *Adobe Support* website

```
http://kb2.adobe.com/cps/311/311742.html
```

has many useful hints on how to create more efficient vector files, as well as information on the influence that specific graphics effects and elements have on memory requirements, and on potential problems when editing vector files. The performance of graphics software when editing a large vector file can, for example, be significantly improved by splitting complex outlines with many anchor points or many curves, into shorter segments. Interestingly, MATLAB seems to have already carried out this type of polygon segmentation, as is suggested by the *coastline_linegraph_vs1_matlab_17703vertices.eps* file. If you prefer continuous polygons that have not been broken up into segments, it would be better to export filled polygons from MATLAB and change the fill color to white after importing the file into the graphics software, as described in Section 6.6. MATLAB appears to choose the *Split Long Paths* option when writing an EPS file, in order to save memory. In contrast, zones that are filled with a uniform color are shown as closed polygons and are saved in a different layer at a lower level. Merging segments in the manner described in the subsection on line graphs can be a time-consuming task that you may wish to avoid.

Let us now open *coastline_linegraph_vs1_matlab_8852vertices.eps*, with 8,852 vertices, in our graphics software. Unfortunately, *Inkscape* seems to be completely unable to open large vector files, as indicated by the error message *Failed to load the requested file*. Opening a PDF version of the same document seems to be possible, but the overall performance of the software is poor as refreshing the graph after introducing any modifications is extremely slow. We can consult the *XML Editor* panel to view the list of coordinates for all of the vertices, as an example of a behind-the-scenes look at an *SVG editor*. The commercial Illustrator software, in contrast, is very good for editing the *coastline_linegraph_vs1_matlab_8852vertices.eps* file. As expected, we find that the black contour paths have been split into shorter segments, apparently in order to optimize file performance, and they are placed into a separate layer.

Pseudocolor Plots: Combining Raster with Vector

The pseudocolor plot created in Section 6.3 depicts the ETOPO2 digital topography and bathymetry on a regular 2-minute grid. Opening the file *etopo2_pseudocolorplot_vs1_matlab.eps* reveals that MATLAB generated a raster file instead of a vector file, and in so doing lost the vector information for the axes, ticks and labels. There are several ways around the problem of rasterizing a vector graph, including a pseudocolor layer. The first alternative is to create a raster version of the graph with disabled axes, ticks, and labels using the *axis off* command in MATLAB, to save this as a first version of the graph, and then to create an empty set of axes with only the axes, ticks, and labels and save this as a second version. Both files can then be combined using vector graphics software such as those described above, rather than having to redraw or retype the vector objects of the graph.

A second alternative is to create a new layer above the layer containing the original pseudocolor plot and then redraw and retype the axes in the graphics software. We can also crop the image by creating a *clipping path* in our vector software. Alternatively, we can crop the colored interior of the graph using pixel software such as *Gimp* or *Adobe Photoshop*, save it as a JPEG or TIFF file, place this file into the vector file and save the illustration as *etopo2_pseudocolorplot_vs3_ai.eps*.

Block Diagrams: Three-Dimensional Visualization and Labeling

The ETOPO2 data set, as created in Section 6.3 and stored in the *etopo2_surfaceplotlight_vs1_matlab.jpg* file, shows an attractive, detailed, and fully colored three-dimensional digital elevation model. We will refine this graphic by adding labels and presenting it as a three-dimensional block diagram (Fig. 8.5).

We first open the original image in Photoshop and reduce the document size to a width of 2,700 pixels only, which means that a 20 cm wide printout has a resolution of 300 dpi. We then cut the white background and save graphic as *etopo2_surfaceplotlight_vs2_ps_cropped.jpg*. We now create a new page 100 mm square using *Adobe Illustrator* and emplace the cropped image using *Menu Bar > File > Place … *. If the cropped image has a

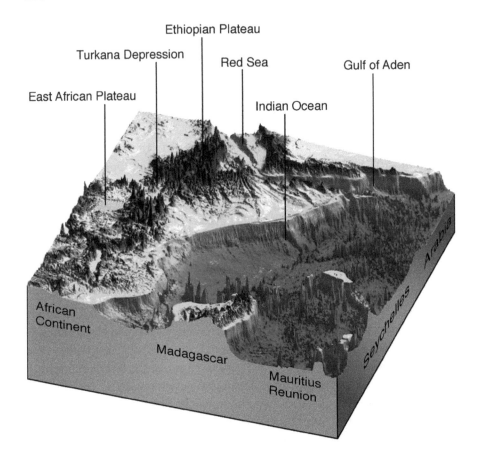

Fig. 8.5 Block Diagram: Three-Dimensional Visualization and Labeling.

white frame around the actual graphic and is therefore slightly larger than 100 mm square we can carefully adjust the dimensions of the image by holding the *Shift* key down and the image will be automatically adjusted to fit the page. We save the preliminary version of the graphic in a file that we call *etopo2_surfaceplotlight_vs4_ai_raw.eps.*

We next create some labels to indicate important locations on the map. To do this we first create a new layer called *text,* choose *Toolbar > Type Tool* and type *East African Plateau.* We mark the text object using the *Character* panel, define *Helvetica Regular 8 pt* as the font type and size, and then drag the text frame into the correct position. To save time we duplicate this text object several times using *Copy & Paste,* or alternatively, by dragging the text frame to another position while keeping the *Alt* key depressed. We then change the text of the various text objects to *Turkana Depression,*

Red Sea, Indian Ocean and *Gulf of Aden*. Below the 3D surface, we rotate the *African Continent, Madagascar* and *Mauritius Reunion* labels using *Menu Bar > Object > Transform > Rotate*, try out and choose an *Angle* of –11°. We likewise type and rotate the *Seychelles* and *Arabia* labels, choose an *Angle* of 59°, and move all of the labels to their appropriate positions.

Then, we create a new layer called *box* for the three-dimensional block underneath the surface plot. We then select *Toolbar > Line Segment Tool*, press the *Shift* key, and draw a vertical black line 21 mm high with a *Stroke Weight* of 0.6 pt, which we move to the left edge of the graph. We make another two copies of this object, change the height to 8.6 mm for the edge on the right and to 13.7 mm for the edge in the foreground. We then create a horizontal line 73.6 mm long, rotate it to –11° as for the labels above, and connect the bottom of the vertical line on the left with the bottom of the line in the center. A line 54 mm long and rotated by 59° connects the bottom of the vertical line in the center with the bottom of the line on the right.

Unfortunately, the lines defining the 3-D block are not yet visible since the *box* layer underlies the opaque *bitmap* layer containing the colored JPEG image that we placed there previously. In the next step of our editing process we will eliminate the white background. We find the file name *etopo2_surfaceplotlight_vs2_ps_cropped.jpg* written in blue on the left side of the *Context Menu Bar*. Clicking on this file name opens a second *Context menu*, in which we choose *Edit Original* to open the original file in our preferred image processing software, which in our case is Adobe *Photoshop*. Using *Menu Bar > Select > Color Range*, in Photoshop we select the white color and set the *Fuzziness* to 100, which affects the falloff beyond the selected boundaries. In the *Layers* panel, we click on the only layer present, which is called *Background,* and rename it as *Layer 0*. Using the *Backspace* key deletes the white zone and in its place we get a transparent zone around the image instead of the former opaque white background. Since JPEG files do not allow the use of transparency, we save the file as a PNG file called *etopo2_surfaceplotlight_vs5_ps.png* and return to Illustrator.

Using *Context Menu Bar > Linked File* and then clicking on the name of our original JPEG image and the small *Relink* icon at the bottom left, we establish a new link. While doing so, we replace the previously emplaced JPEG image with our new PNG file. Using *Context Menu Bar > Embed* we then embed the PNG file into the *figure_2e_chapter_6_edit_vs2.eps* file. In the *Layers* panel, we create a new layer called *strokes*, which we move together with the *text* layer to a position above *Layer 1* to ensure that they always remain visible. In the *strokes* layer we then connect the upper labels with the corresponding zones of the illustration by drawing vertical lines of

0.6 pt thickness. We then save the file as *etopo2_surfaceplotlight_vs6_ai_labels.eps*.

Let us again explore some advanced applications of the graphics software, starting with the use of color gradations. We first create a new layer called *box color zone* at the bottom of the list of layers, for the color of the vertical surfaces of the block diagram. We then choose *Toolbar > Pen Tool*, select an arbitrary sandy, tan or grey color from the *Swatches* panel for the fill, and draw the south and east facing vertical surfaces of the block diagram. As the *box color zone* layer is below *Layer 1* and the PNG file contains an image with no background, we can now draw the upper line of the rectangle freehand, we only have to ensure that we meet the corner points of the existing vertical and diagonal block outline. We then select one object from the *box color zone* layer, choose the purple radial gradient from the *Swatches* panel for the fill, and open the *Gradient* panel to adjust the gradient. We find the gradient displayed as a vertical guide comprising two rulers, with a white and a purple color swatch below the guide. We click the white swatch and move it to the left to adjust the gradient. Next we choose the pale tan color C = 25, M = 25, Y = 40, K = 0 from the *Swatches* panel and use it to replace the white color in the gradient ruler of the *Gradient* panel using *Drag & Drop*. Similarly, we replace the purple color in the *Gradient* panel with beige (C = 35, M = 60, Y = 80, K = 25), and using *Drag & Drop* we select the new gradient in the *Toolbar* and add it to the *Swatches* panel. In the event of problems occurring while using the illustration in a subsequent application, such as text processing, or printing it out, we can isolate and export the critical gradient elements into a bitmap file and then later replace them in the illustration, thus keeping the text and contours as vectors in the same way that we demonstrated for the transparent zones of the pie chart. We save our file under the name *etopo2_surfaceplotlight_vs7_ai_block.eps*.

As a second advanced application, we will now work on the representation in perspective of the inclined labels on the block diagram. We first shear the three labels *African Continent*, *Madagascar* and *Mauritius Reunion* using *Toolbar > Object > Transform > Shear*, and then type *Shear Angle −11°* and *Axis Angle 11°*. Please ensure that the *Preview* checkbox is activated to permit inspection of the final result. We then shear the *Seychelles* and *Arabia* labels, choosing 59° for both *Shear Angle* and *Axis Angle*. To improve the appearance of the labels, we open the *Character* panel and type 150 % for the horizontal and 125 % for the vertical alignment of the letters. Finally, we move the inclined labels to appropriate positions, set black strokes and text overprinting using the *Attributes* panel, and save our final graphic as *etopo2_surfaceplotlight_vs8_ai_perspective.eps*. In the event of file problems in subsequent applications due to color gradation, we can replace the zone containing color gradation, by exporting it as a bitmap file and and again placing it in an Illustrator document, as described above for the pie chart. We then save the file as *etopo2_surfaceplotlight_vs10_ai_flattened.eps*.

Preparing Vector Graphics for Presentation on a Dark Background

Some users may wish to present vector graphics on a dark background, for a poster or presentation, as will be demonstrated in Chapters 9 and 10. To demonstrate how to modify line drawings for that purpose, we open two plots, both of which have been edited in this chapter: the pie chart stored in the *icecore_piechart_vs11_ai_mask.eps* file and the block diagram in the *etopo2_surfaceplotlight_vs10_ai_flattened.eps* file. We first change all black fonts to white in both graphics, and then change the black outlines in the second graphic to white. Finally, we deactivate the overprinting checkbox for white contours and fonts in the *Attributes* panel. We save versions 11 and 12 of the graphic with file names ending in *...fordarkbackground.eps*. To prepare a third graph with dark background for inclusion in a poster or presentation, we open *icecore_lineplot_vs4_ai.eps*, change all font colors to white, add a small square in the same color as the curves to link the curves with the axes, and save the file as version 8 with a file name ending in *..._fordarkbackground.eps*.

8.3 Processing Images

In Chapter 7.4 we imported and processed an ASTER satellite image and saved the result as a TIFF-file called *naivasha.tif*. We now prepare this image for inclusion in conference presentations, posters, or manuscripts. Typing whos in MATLAB indicates that the image is stored as a $4200 \times 4100 \times 3$ array, consisting of three 4200×4100 arrays, one for each of the colors red, green and blue (see Chapter 7.4). The listing of the current variables in the workspace also provides the information that it is a *uint8* array, i.e. each array element representing one pixel contains 8-bit integers. The size of the image in the workspace is $4200 \times 4100 \times 3 \times 8$ bits = 41,3280,000 bits (b), or 41,3280,000 bits / 8 = 51,660,000 bytes (B), or 51,660,000 bytes / 1024 = 50,449 kilobytes (kB), or 50,449 kB / 1024 ≈ 49.3 megabytes (MB). The uncompressed baseline TIFF-file created by imwrite in MATLAB is slightly larger (52.1 MB) since it includes a header providing information on the size of the image, compression method (if any), and other properties of the image.

The satellite image processed with MATLAB is very poor in contrast, which needs to be fixed before printing. The Image Processing Toolbox in MATLAB software provides numerous methods for image enhancement, such as adaptive or non-adaptive histogram equalization using `histeq` or `adapthisteq`, respectively. Nevertheless, if the image will not be used for quantitative analysis but is to be printed out for use in fieldwork, any image processing software (such as *Gimp*, *GraphicConverter*, or *Adobe Photoshop*) can be used to quickly enhance the image.

Satellite Imagery: Editing a Processed Satellite Image using Gimp

We use the GNU Image Manipulation Program (*Gimp*) freeware to process the satellite image. After launching the software, we use *Open* from the *File* menu to navigate to the directory in which we have stored the *naivasha_image_vs1_matlab.tif* file and open the file. A small window pops up stating that there is no color profile defined and requesting us to choose either an RGB or a CMYK profile. When processing an image we always use an RGB color model such as, for example, the sRGB IEC61966-2.1 color profile, which is the standard profile for digital cameras. We then obtain an image that looks very similar to the one that we saw in MATLAB. We can read the header of the file, or the actual image, using *Image Properties* from the *Image* menu. According to the header, the print size of the image is 1446.39 × 1481.67 millimeters or almost 1.5 × 1.5 meters. This size results from the 72 × 72 pixels per inch (ppi) resolution of the image and the simple calculation that, since an inch is 25.4 mm, 72 pixels per inch corresponds to about 2.8346 pixels per millimeter; the 4200 × 4100 pixel image therefore has a size of 1446.39 × 1481.67 millimeters.

This print size for the image is too large to be included in a manuscript, presentation, or poster. On the other hand, the resolution of 72 ppi is relatively low compared to the standard resolution of printed magazines which is 300 ppi, this being the limit that the unaided human eye can differentiate at a typical reading distance. Scaling the image to 300 ppi using *Scale Image ...* from the *Image* menu decreases the print size to 347.13 × 355.60 mm, or about 35 × 35 cm, without interpolating or reducing the number of pixels. The size of the image, however, is still too large for most printed documents. We again use *Scale Image ...* to reduce the width of the image to 150 mm. The lock icon to the right of the text input locks the proportions of the image and calculates a corresponding height of 153.67 mm, since the image is not quite square. Keeping the 300 ppi resolution, the software now reduces the file size of the image using an interpolation method of the user's choice, e.g., a linear interpolation technique. We save the resulting image onto our hard drive without compression, as a 9.7 MB file called *naivasha_image_ vs2_gimp_1772px.tif*.

We next explore some of the image enhancement features available in the Gimp software. As with most image processing software tools, Gimp provides automated color and contrast enhancement tools. For example, we can use *White Balance* from the *Colors > Auto* menu. According to the Gimp manual, the *White Balance* command automatically adjusts the colors of the image by stretching the red, green and blue channels individually. To do this, it discards pixel colors at each end of the red, green and blue histograms, which are used by only 0.05 % of the pixels in the image, and stretches the remaining range as much as possible. The resulting satellite image indeed looks much better and we save it onto our hard drive without compression as a 9.7 MB file with the name *naivasha_image_vs3_gimp_ autocolor.tif*. The software provides many more image enhancement methods that can be applied in other examples.

The 9.7 MB, 150 × 150 mm image, with a resolution of 300 ppi, is rather large, especially if it has to be sent by email or included in a webpage. We can use image compression to reduce the file size. The most popular 24-bit true color file format and compression method is the JPEG method, with the name being an acronym for *Joint Photographic Experts Group*, which is the name of the team that invented the method in 1986. This compression method involves grouping pixel values into 8 × 8 blocks and transforming each block with a discrete cosine transform. All unnecessary high-frequency information is consequently deleted, which makes this compression method irreversible. As an example, let us save our enhanced image in the JPEG format using *Save Image …* from the *File* menu. The dialog window again pops up and allows us to select the *JPEG image* file type, with 70 % quality. We save the file as *naivasha_image_vs4_gimp_smallfile.jpg*. Reopening the file reveals the typical JPEG artifacts including the 8 × 8 blocks and the loss of small-scale detail of the image. However, these artifacts are only visible at the 800 % zoom level and the file size has been reduced from 9.7 MB to just 0.618 MB (or 618 KB). Viewing both versions of the image at 100 % zoom level does not reveal any differences.

Satellite Imagery: Image Processing using Adobe Photoshop

The *Adobe Photoshop* professional software tool included in the *Adobe Creative Suite* provides a commercial alternative to *Gimp* that is a powerful tool for editing images. Launching the program, the graphical user interface (GUI) is similar to that in the companion product *Adobe Illustrator* that we used previously, comprising the typical Adobe menu bar, toolbox, and panels. Additional information and excellent video tutorials are available from *Adobe Help* and online at

```
http://help.adobe.com/en_US/photoshop/cs/using/index.html
```

or

```
http://tv.adobe.com/
```

To provide a comparison with *Gimp*, we work through a similar process with Photoshop. Using *Menu Bar > File > Open*, we open the file called *naivasha_image_vs1_matlab.tif* and use *Menu Bar > Workspace > Photography* to choose the workspace, which shows a preselected set of panels. The *History* panel shows a list of working steps that allow us to go back and forth within the program. We adjust the image size using *Menu Bar > Image > Image size*, to a resolution of 300 dpi and a width of 2,500 pixels.

The *Layers* panel shows our image within the layer called *background*, which is locked, as indicated by a padlock symbol. This offers the possibility of *non-destructive editing* while keeping the original image untouched. We therefore use a separate *adjustment layer* for color correction, making use of the *Curve* tool icon from the *Adjustments* panel. Clicking on the *Auto* button changes the color gradation curves for the three (RGB) color channels, and creates a new layer in the *Layers* panel, with enhanced colors and contrast. We save version 5 of the file as a Photoshop file with the *.psd* extension. As files with Photoshop formats are not always compatible with later software, we also save the file as *naivasha_image_vs6_ps_rgb_2500px.jpg*, with the color profile *sRGB IEC61966-2.1* embedded. Alternative strategies for color adjustment are provided by the *auto-contrast* and *auto-color* options of the software, which can be used to enhance an image (Fig. 8.6) as well as for direct adjustment of the *gradation curve*.

As an advanced exercise in this subsection we will now prepare our image for offset printing, using the four CMYK ink colors. A special feature of the Adobe Creative Suite is the *Color Management Workflow*, which can be controlled and harmonized in the *Adobe Bridge* using *Menu Bar > Edit > Creative Suite Color Settings*. The requirements for a perfect color management workflow include a professionally calibrated computer display, software, various technical tools, and the ability to simulate the printed outcome on the screen prior to actual printing. We continue with version 6 of the image, open the *Channels* panel, and after toggling the visibility, we can

Fig. 8.6 Automatic color adjustments - from the left to the right: **a** Gimp auto white point balance, **b** Photoshop adjustment auto gradation, **c** Photoshop adjustment auto contrast, **d** Photoshop adjustment auto color.

see the three RGB channels used to present the image on a computer screen. As the image is intended for offset printing using the four-color process, we change the color profile into a CMYK color profile using *Menu Bar > Edit > Convert to profile.* In this example, we embedded the *Coated FOGRA27* color profile into the file resulting in a 7.6 MB file, compared to the smaller 3.3 MB RGB file of version 6 with the same image dimensions, resolution, and compression. In the corresponding dialog box, we choose a suitable color profile from the *Destination Space* drop-down menu. Activating the *Preview* checkbox allows the appearance to be checked, testing the four different *Intent* options for color conversion. In the *Channels* panel, we now find the four printing colors cyan, magenta, yellow, and black (also called "Key"). We save the new file as *naivasha_image_vs7_ps_cmyk_2500px.jpg*, with high quality compression. More information about colors is provided in the section on *Colors and Color Management* in Chapter 11.

Georeferenced Satellite Imagery: Masking, Retouching, Adding Vectors

In Section 7.5 we georeferenced the satellite image, i.e. the image was rotated, scaled, and deformed to fit a geographic coordinate system. We then fur-

EDITING GRAPHICS, TEXT, AND TABLES

8

ther edited the image in the preceding subsection of this chapter, by auto-adjusting the colors and contrast, and converting the color model from RGB to CMYK. The resulting raster image lies on a black background, shows a raster coordinate system, and labels. Our task in this subsection will be to enhance the colors and contrast, to remove the background, and to add a vector coordinate system, labels, scales, and vector objects such as tectonic faultlines and the outlines of waterbodies. The original georeferenced version in *naivasha_georef_vs1_matlab.tif*, generated by MATLAB, is a very large file (216 MB) and the image is 8,570 pixels wide, resulting in a width of 72 cm at 300 dpi resolution. To reduce the file size we choose *Menu Bar > Image Size*, where we can set the *Resolution* to 300 dpi and reduce the *Width* to 2,500 pixels. Activating the *Chain* symbol ensures that we keep the original proportions of the image. We now save the 2.2 MB file as *naivasha_georef_vs3_ps_rgb_2500px.jpg*.

To select and remove the black background of the image, we now choose the *Lasso Marquee* tool and click on the four corner points of the colored area to make a selection around the edge of the satellite image. We can press the *Alt* key to reduce the size of the selected area, or the *Shift* key to increase it. To select the black zone we then invert our selection using *Menu Bar > Select > Inverse*. Using the *Backspace* key opens the *Content* dialog window and we choose to replace black with white. To control the selection, we can always switch view between the *Standard* mode and the *Mask* mode by toggling the lowermost tool in the toolbox. In *Standard* mode, the borders of the selection are marked by an animated border resembling crawling ants. In the *Mask* mode, the selection is indicated by a red semi-transparent zone overlying the image. In case we want to reload the selection later, we can save the mask as an *alpha channel* by using *Menu Bar > Select > Save Selection*. Then, because the JPEG file format does not allow us to save separate layers, in order to retain the alpha channel, we save the image as a Photoshop file named version 3. We also save the image as a JPEG file named *naivasha_georef_vs4_ps_rgb_white.jpg*. When adjusting the image colors (with either a black or a white background) using *Menu Bar > Image > Adjustments > Curves > Auto*, the results are much less attractive in terms of color intensity, brightness, and contrast than in the non-georeferenced version. This may be a result of the process taking into account the black or white background.

For a georeferenced image combined with a coordinate system, we open *naivasha_georef_vs2_matlab.jpg* and try to again adjust the colors using *Menu Bar > Image > Adjustments > Curves > Auto*. For some reason this has almost no effect on the colors. We could of course just copy and paste the version 3 image generated above. Let us assume, however, that we wish to use the non-georeferenced image stored in the file called *naivasha_image_vs6_ps_rgb_2500px.jpg* as it has nice colors, and that we want to georeference this image using *Photoshop*. If the rulers are not visible in the document window we can make them active using *View > Rulers*. By clicking and dragging the mouse down from the horizontal *Ruler*, we create two horizontal *Guides* and, similarly, two vertical *Guides* from the vertical ruler. With the aid of the *Guides*, we mark the four corner points of the satellite image. We now copy and paste the image from *naivasha_image_vs6_ps_rgb_2500px.jpg* into a new layer at the top of the layer list, and can then observe the image while we transform it into a trapezoid using *Menu Bar > Transform > Distort*. Finally, we retouch the black background behind the image by drawing a white rectangle between the guides in a layer between the two images, and we then save the file in the Photoshop file format as version 5, or alternatively as a JEPG file called *naivasha_georef_vs6_ps_rgb_coord.jpg*.

To demonstrate how to add vector objects in Photoshop, we now add text and some contours using the *Pen* or the *Line Drawing Tool*, and the *Horizontal Type Tool* from the toolbar. The tools look very similar to the Illustrator tools and we can again organize all different types of elements into separate layers. To demonstrate the limitations in compatibility between different modules of the Adobe Creative Suite, we open and then copy some random line-drawing elements. For example, we can select and copy the *fluvial system* layer included in the file named *srtm_faults_sat_merged_vs1_ai.eps* from Illustrator onto the computer clipboard, paste it into our Photoshop file as a *Smart Object*, and adjust its dimensions. Unfortunately, before we can continue with our work the program tells us that all objects have now been converted into a pixel format, as indicated by the message *Rendering Placed Document*.

For text and line drawings, we prefer to preserve editable vector text, line drawings, and elements rather then having them rasterized by *Photoshop*. To retain vectors we therefore follow a different strategy and go back to using *Illustrator*. We reopen the vector file *srtm_faults_sat_merged_vs1_ai.eps* using *Menu Bar > File > Place ...* and import the raster image file called *naivasha_georef_vs6_ps_rgb_coord.jpg* into the *sat coordinates* layer, and then adjust its size and position there to create congruency between both coordinate systems. We can type labels for *Lake Naivasha*, *Mau Escarpment* and *Aberdare Ranges* in the uppermost layer, as we did with Photoshop. When saving the illustration, we observe the file size and discover that the *.eps* vector file that was originally 1.2 MB becomes a very

large file when combined with the original georeferenced raster image. A
.ai file with the option to embed a PDF deactivated will have a smaller file
size than the *.eps* file, but for reasons of compatibility we save it as a *.pdf* file.
To remove the white background of the image and reduce the file size, we
replace the image with a version that has no background called *naivasha_
georef_vs9_ps_cmyk_670px.pdf*, generated in Photoshop, and finally save
our illustration as *srtm_faults_sat_merged_vs3_ai.pdf*.

8.4   Editing Text

This section is on preparing text for integration into manuscripts, posters
and, less commonly, into presentations. Word processing is one of the oldest applications of computers and the emergence of sophisticated software
tools that were originally independent from each other and each of which
subsequently claimed to be market leaders, has led to a wide variety of text
formats and consequent incompatibilities between operating systems and
software tools. The *Guidelines for Authors* of most journals therefore recommend that the layout of a manuscript be kept simple, avoiding the software's options to justify text or hyphenate words, and that the manuscript
instead be submitted as single-column left-justified text.

The text format with the highest possible compatibility is the *American
Standard Code for Information Interchange* (ASCII) format, which was
first published in 1963, by the American Standards Association (ASA). As a
7-bit code, ASCII consists of $2^7 = 128$ characters (codes 0 to 127). Whereas
ASCII-1963 lacked lower-case letters, these were subsequently included in
the ASCII-1967 update, as well as various control characters such as *escape* and *line feed*, and various symbols such as brackets and mathematical
operators. Since then, a number of variants have appeared to facilitate the
exchange of text written in non-English languages, such as the expanded
ASCII, which contains 255 codes, e.g., the Latin-1 encoding.

One of the classic examples of cross-operating system or software incompatibilities is the *carriage return/line feed* problem. *Carriage return*
(CR) in computing refers back to the return of the carriage on a typewriter,
and is an ASCII control character to move the curser back to the first column, or position, on the same line. *Line feed* (LF) moves the cursor to the
next line on a typewriter, or starts a new line on a computer. Both are used
together to mark the end of a paragraph and the start of a new one in most
word processors. Unfortunately, some software tools use the carriage return character, and some only the return character, to mark the end of a
paragraph, while other software tools use carriage return together with line

feed. The problem is that text generated on Unix-based operating systems such as Linux, SUN Solaris, or Mac OS X using LF alone as end marks for paragraphs appear as a single long line on Windows computers. On the other hand, Windows-sourced files with both CR and LF appear with a second paragraph break and hence there is an extra line between paragraphs. Although most word processors can now deal with this problem, users still come across the CR/LF problem when working with low-level word processors such as those used for coding software.

The more advanced and feature-rich the word processor is, the more incompatibilities appear when exchanging text between different computer platforms or software. In 1987 the Microsoft Corporation introduced the cross-platform 8-bit *Rich Text Format* (RTF), which includes *escape sequences*, allowing the text to be formatted using a *WYSIWYG* (*what you see is what you get*) type text processor (http://microsoft.com). Binary DOC files generated by *Microsoft Word*, ODT files generated by *OpenOffice Text*, and *Pages* files generated by *Apple Pages*, all contain a great deal more formatting information than RTF files, but at the expense of being less compatible. *Apple Pages*, however, contains converters to read DOC files and convert them into *Pages* files, but the latest 2011 release of Microsoft Word can not read *Pages* files.

When formatting a text in a text editor such as *Word*, *Pages* or *OpenOffice*, the application generates a code showing a typical syntax, similar to that used in other programming languages. Well-known examples of coding, tagging coding, and tagging texts are included in the HTML or XHTML mark-up languages that are used to create webpages. When writing a piece of text, all text elements are tagged using a structure such as *header1*, *header2*, *header3*, *normal*, *indent*, *table*, or *list*, which enables the use of *styles* instead of directly formatting each individual character and paragraph of a manuscript. The characteristics of a *paragraph style*, such as its *font*, *font weight*, *font size*, *color*, etc., are defined in a central document called a *style sheet* and can be administrated therein. The formatting of individual characters, individual letters, or parts of text within a paragraph, is defined in the *character style*, which is then available throughout the entire document. For websites, this document is called a *css* or *cascading style sheet*, which is used to improve the appearance of the site. Storing all style information in a central place allows different styles to be applied and, if necessary, changed quickly, consistently, and safely throughout the entire document. If such a structure of *paragraph styles* and *character styles* has been applied from the very beginning it will facilitate subsequent handling of the document when, e.g., generating the table of contents or editing the

EDITING GRAPHICS, TEXT, AND TABLES

8

text, and will allow cross-platform use for many different purposes. The easiest way to get started with the use of styles is to adopt a ready-to-use template, such as may be provided by the program itself, or downloaded either from the authors' guidelines of the journal in which it is to be published, or from the institution's graphics design department we want to publish for.

Starting *Microsoft Word* on a *Mac* computer opens a *Word Document Gallery* with ready-to-use templates for many purposes, as well as any personal templates in the *My Templates* folder. We choose *File > New Blank Document* and then *View > Toolbars > Standard*, followed by *Formatting* and the *Styles* toolbox, which opens the vertical *Menu Bar*, the *Standard Toolbar*, a floating *Toolbox* and the *Ribbon* with index card tabs. Clicking on *Menu Bar > Word > Preferences* opens a pop-up menu in which we can personalize the program with respect to *authoring*, *proofing*, *output*, *sharing* and *personal settings*, for example, by changing the *user information* and *autocorrection settings*.

We type some placeholder text (for example: *Heading 1*, *Heading 2*, *Heading 3*, *Text Hello World*) and format it using the paragraph styles with the same names presented below, or using *Pick a Style to Apply* or by choosing a style from the *Styles* menu in the *Ribbon*. Hovering over a style in the *Styles* menu shows the format details, i.e. the *Font*, *Font Color*, *Space*, *Numbering* and *Theme*. In the *Ribbon* we can choose different themes, which are built-in sets of fonts and colors. The *Office* theme is suitable for a scientific report, using the serif font *Cambria* for the main text and the non-serif *Calibri* for the headings. Clicking on *Menu Bar > Format > Font*, or *Paragraph*, or *Document*, opens a pop-up menu with formatting options. A similar but more compact menu with a larger number of options opens when we click on a style in the pending *Styles* palette and choose *Modify Style* in the *Context* menu. In the lower left corner of the drop-down menu we can choose options such as *Font*, *Paragraph*, *Tabs*, and *Numbering*.

We need some more text and search the web for any *blindtext* generator, and then paste some 200 words as running text below each heading. We click on the *Document Elements* index card tab of the *Ribbon*, insert a fully formatted placeholder for a *cover page*, a *header*, and a *table of contents*, and then save our document as *report_word.docx* or *report_word* in any other file format. To increase efficiency and to maintain an overview of invisible elements in a document, we choose *Show all Nonprinting Characters* in the toolbar, which shows *page breaks*, *paragraph marks* or *pilcrows* (¶), and *soft returns* or *manual line breaks* (↵). Each end of a paragraph requires a paragraph mark, which also embodies the paragraph style. To check the formatting at a glance, we click on the *Show Styles*

Guides in Document checkbox in the pending *Styles* palette. Finally, having formatted the table of contents, we save the file as a *Word Document* and also, for future use, as a *Word Template*. When opening a text from another computer the formatting may look a little strange if some of the fonts are missing. Word substitutes missing fonts without any warning, but this can be adjusted in the *Styles* palette.

We now continue editing the same document but using *Apple Pages*. Opening *Pages* for the first time shows a welcome window, a brief getting started video tutorial at

```
http://apple.com/iwork/
```

followed by the *Template Chooser* window with templates for word processing and page layout. We explore the menu bar and then open the *MS Word* file *report_word_nonserif.docx*. In *Pages*, a *Document Warning* window might then open in which we can replace any missing fonts. To customize our workspace we open the *Styles Drawer* by activating it from *Menu Bar > View > ...* . There we can make visible *Invisibles, Comments, Rulers,* and also the main tool for formatting, the *Inspector* window, too. In the *Styles Drawer* we find a list of the same styles as were defined in Microsoft Word, as well as a preset of style offered by *Pages*. To demonstrate how changes can be made very quickly, intuitively and consistently, we place the cursor within *1. Heading 1* in our text, open the *Colors* window, and change the color to black for some letters of *Heading 1*. After clicking on the small red triangle in the *Styles Drawer,* we first choose *Select All Uses of heading 1*, and then *Redefine Style from Selection*. We decide to delete the existing styles and instead use the default styles. To achieve this we click the cursor within the text *Heading 1*, choose *Select All Uses of heading 1*, and click on the *Heading 1* without numbering style format offered by the program. To add automatic numbering, we use *Text > List > Bullets and Numbering* and *Indent Level* from the *Inspector* palette. For pagination, we click on the header, choose *Menu Bar > Insert > Auto Page Numbers* and adjust the options in the *Insert Page Numbers* menu. Using *Text > More > Borders and Rules* in the *Inspector* palette we create a separator line of 0.5 pt thickness between the header and the main text. We again transfer this *Direct Formatting* to the *header 1* style by clicking on the red triangle in the *Styles Drawer,* which indicates a discrepancy between the old an the new formats, and we choose *Redefine Style from Selection*.

As an example of adding a figure, we simply click on *Menu Bar > Insert > Choose* and import *icecore_piechart_vs11_ai_mask.eps*, the pie chart edited in Section 8.2. Using *Menu Bar > Format > Mask with Shape > Rectangle* and *Edit Mask*, we can crop the white space around the image and adjust its size. Finally, we save it as *report_pages_nonserif.pages*.

8.5 Editing Tables

In this section we deal with preparing tables for integration into manuscripts and, less commonly, into posters. Tables should not generally be included in presentations. As with word processing, tables were an early application of computers since the first computers were designed to process data rather than text, and data are usually stored in tables with rows and columns.

In Chapters 4 to 7 we used MATLAB to import, process, and visualize data organized as arrays of numerical values, as well as text and other types of data. We have not used Microsoft Excel spreadsheets, even though MATLAB can read its files using `xlsread`. Instead, we mostly used `load` to import data from ASCII files, and in some rare cases we also used other tools to read specific binary formats such as images or digital elevation models. The reason for using ASCII instead of Microsoft Excel files is to avoid the incompatibilities that we often come across when working with binaries. Again, the *Guidelines for Authors* for most journals recommend that the layout of tables attached to a manuscript be kept simple. Both tables and figures are generally submitted as appendices to a document manuscript, rather than integrated into the text.

Tables are typically two-dimensional rectangular arrays of numbers or text, organized in rows and columns. In contrast to data arrays to be imported into MATLAB, tables to be included in printed manuscripts are restricted in their dimensions by the page format. The widths of the tables in this book, for example, are limited to the 11.69 cm width of the text column. As an example, we use the text file called *geochem_data.txt*, which was saved as an ASCII text file in Chapter 4.

```
SampleID    Percent C       Percent S
101           0.3657          0.0636
102           0.2208          0.1135
103           0.5353          0.5191
104           0.5009          0.5216
105           0.5415         -999
106           0.501          -999
```

This file is in a very common format for tables downloaded from internet resources. The first row contains the names of the variables and the columns provide the data for each sample, i.e. the sample identifications, and the percentage of carbon and sulfur in each sample. The absurd value *-999* indicates missing data in the data set. The columns of tables are separated by *delimiter characters*, commonly single spaces, multiple spaces, or tabs. In some examples, commas, colons or vertical bars (or pipes) are also used as delimiters. The lines are separated by *newlines*, or *line breaks*, or *end-of-line* (EOL) markers, which in ASCII character sets are commonly CR, LF or CR/LF (see Section 8.4).

Taking a closer look at our example file, you may notice that the delimiters of the values are multiple spaces, which is the worst-case scenario for desktop publishing. We can *unhide* the control character used by our word processing software in order to reveal the spaces:

```
SampleID···Percent·C·······Percent·S
101··········0.3657··········0.0636
102··········0.2208··········0.1135
103··········0.5353··········0.5191
104··········0.5009··········0.5216
105··········0.5415··········-999
106··········0.501···········-999
```

Formatting such a table is a nightmare and we therefore need to replace the spaces by tabs. Tabs, or tab stops, on a typewriter halt the carriage movement by mechanical gears and allow text to be aligned vertically. In word processing, the same idea is used to align text with the left and/or right margins, to centralize it, or to arrange numerical values in such a way that the decimal points in figures on separate lines are positioned directly beneath each other.

There are a number of different ways to replace the multiple spaces with single tabs. One way is to first replace double spaces with single spaces using the *search-and-replace* feature of the word processor. We use this feature repeatedly until no more double spaces are found.

```
SampleID·Percent·C·Percent·S
101·0.3657·0.0636
102·0.2208·0.1135
103·0.5353·0.5191
104·0.5009·0.5216
105·0.5415·-999
106·0.501·-999
```

We can then replace single spaces with tabs, again using *search-and-replace*. The control character for a tab is a small arrow. Some of the tabs mistakenly inserted in the header have to be changed back to spaces.

```
SampleID→Percent→C→Percent→S
101→0.3657→0.0636
102→0.2208→0.1135
```

```
103→0.5353→0.5191
104→0.5009→0.5216
105→0.5415→-999
106→0.501→-999
```

We can then change the location of the tab stops to alter the layout of the table in the word processor, e.g., by introducing two left tab stops, at 3 cm and 6 cm.

```
SampleID→          Percent C→      Percent S
101→               0.3657→         0.0636
102→               0.2208→         0.1135
103→               0.5353→         0.5191
104→               0.5009→         0.5216
105→               0.5415→         -999
106→               0.501→          -999
```

We can now convert the tab-delimited text into a formatted table, either using the word processor's *table* feature, or by importing the text into spreadsheet software such as Microsoft Excel, OpenOffice Spreadsheet or Apple Numbers. Alternatively, we can import the ASCII text file delimited by multiple spaces into one of the software tools that allow multiple spaces to be replaced directly by single tab delimiters. The binary *.xls* files generated by Microsoft Excel, *.ods* files generated by OpenOffice Spreadsheet, or *.numbers* files generated by Apple Numbers, contain a great deal more formatting information than simple text files, but at the expenses of being less compatible.

The simplest way to import the contents of a tab-delimited table into a text editor or professional *desktop publishing* (DTP) program is to create a new table in the document, and then to select the text and tabs and just *copy and paste* them as a whole into the selected cells. Available formatting options include the possibility of hiding or using visible *cell borders*, choosing the type of fill for the cells in the table, with a choice between different colors, alternating colors, grey or *no fill* for the table cells. Furthermore, we can change the appearance of *headers* or *footers* and adjust the *inset margins* and *text alignment*. In *Pages*, these adjustments are made using *Inspector > Table > Table* and *Format*.

9 Creating Conference Presentations

9.1 Introduction

The results of a project are typically presented in three formats: posters, talks, and papers. A *poster* is collection of graphics, photos, and text printed on a large sheet of paper that is presented on a poster-board in a large hall in a conference building. Before the actual presentation of a poster, the scientist submits an *abstract* summarizing the key findings of the research project. Abstracts are typically limited to between 100 and 200 words although exceptions are sometimes made, e.g., for extended abstracts which, in some cases, may reach the length of a journal article. Abstracts are also required to be submitted for a conference *talk*. A talk is an oral presentation, typically 15 to 20 minutes long and supported by a series of pages or slides projected by a video projector. The slides generally contain graphics and photos, but only a relatively small amount of text. Talks are organized into theme sessions that are chaired by a session convener. The convener introduces the presenter of the talk, takes care of the time management of the session, and moderates the discussion with questions and answers after the talk. This chapter deals with the planning of a talk and the design of presentation slides, as well as offering a few suggestions for practicing and presenting a talk.

9.2 Planning an Oral Presentation

Numerous books, articles and videos already exist on planning an oral presentation, designing the slides, and delivering the talk. Excellent books on oral presentations have been published by Nancy Duarte at *Duarte Design Inc.* (http://duarte.com) and Garr Reynolds (http://garrreynolds.com). Garr Reynolds also has a blog called *Presentation Zen* that provides comments on numerous examples of great presentations given by other people (http://presentationzen.com).

One of the best video examples of presentations is of Steve Jobs, the late CEO of Apple Inc., presenting the first *iPhone* in the Macworld San

Francisco 2007 Conference's Keynote Address, available from

```
http://itunes.apple.com/us/podcast/apple-keynotes/id275834665
```

and from several other websites, including *YouTube* (http://youtube.com). In this video you can see the presentation of the iPhone, which starts at 26:20 minutes into the video and continues for most of the remaining keynote address. The first four minutes or so are actually structured like a scientific presentation. The first sentence, "Every once in a while a revolutionary product comes along that changes everything", clearly builds up a high level of expectation, in particular as there were many rumors about the new mobile phone before its actual launch. In contrast to a scientific presentation, however, Jobs then goes through some of Apple's major achievements and products, rather than discussing the problems with current mobile phones, before then revealing to the audience which new product he is going to talk about over the next hour or so.

At 29:18 minutes he presents the new product by simply displaying its name on a slide (Fig. 9.1). In a science talk or in a paper, this would corre-

Fig. 9.1 Steve Jobs introducing the new iPhone by simply displaying its name on a slide. The sentence, "Apple reinvents the phone", which is written on the slide, clearly indicates the importance of the new product being introduced (Screenshot from the keynote address at the Macworld San Francisco Conference on 1st January 2007, by Apple Inc.).

spond to the *here we present* statement. The statement that "Today, Apple is going to reinvent the phone" clearly indicates the importance of the new product being introduced. Then at 29:48 minutes Jobs presents the *state-of-the-art* of current mobile phones which, in a scientific presentation, would have been outlined before presenting the new product (Fig. 9.2). He does this by explaining a very simple *xy* plot with an *easy-to-use* axis and a *smart* axis. After explaining the axes of the empty graph, he places colored circles on the graph, one by one, representing some of the smartphones offered by competitors, before eventually placing the iPhone in the best possible position on the plot of *smart* versus *easy-to-use*. At 30:58 minutes, he repeats the "Today, Apple is going to reinvent the phone" statement and concludes his introduction of the keynote address before then continuing with an explanation of the technical details and features of the new product for the next hour.

Steve Jobs' presentations have been widely discussed on the World Wide Web by many people, including Garr Reynolds. One of the most striking features of his presentations was the simplicity of the slides, with only a single statement or message per slide. Most slides comprised a single photo

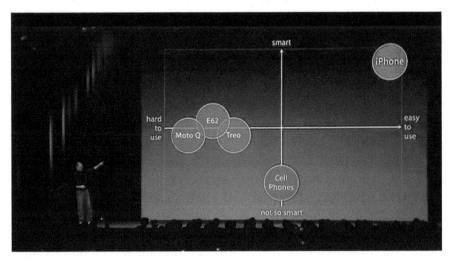

Fig. 9.2 Steve Jobs presents the state-of-the-art for mobile phones, which would be outlined before presenting the new product if it were a scientific presentation. He does this by explaining a very simple *xy* plot with an *easy-to-use* axis and a *smart* axis. After explaining the axes of the empty graph, he places colored circles on the graph, one by one, representing some of the smartphones offered by competitors, before eventually placing the iPhone in the best possible position on the plot of *smart* versus *easy-to-use*. (Screenshot from the keynote address at the Macworld San Francisco Conference on 1st January 2007, by Apple Inc.).

of an item such as a mobile phone, one word such as iPhone, or a very simple graphic. If you go through the entire keynote address, you will find that most of the presentation is made up of still or animated photos of the phone. If text is used then it is reduced to an absolute minimum, such as a short list of the features of a new product, in which all features are described by one or two words only, such as at 1:32:06 hours when the typical features of smartphones are listed. Instead of also listing the features of the iPhone as text, he quickly goes through a series of photos of the phone, showing the features as he talks about them. Then at 1:32:40 hours he presents the two available models by showing photos of the iPhones in the middle and text indicating the size of the flash memory to the left, and the price to the right. Tables are rarely used but are necessary when comparing the features of different products. The tables then have very few rows and columns, perhaps comparing only three to five features in just two products. A good example of how to put together a graphic in a presentation can be found at the end of the presentation (1:41:15 hours). Jobs again uses animation to slowly put together the various elements of a bar plot, rather than straight away displaying the completed plot.

A second point to note in this presentation is the way that Jobs communicated with the audience. In contrast to other presenters, Jobs never showed too much excitement or agitation in his presentations, but he was clearly enthusiastic about the latest Apple products. He also made it very clear several times during his presentations that he worked with a great team. He maintained eye contact with the audience but also admired the products that he and his team had created over recent months and years, when they appeared on the slides. Jokes are well measured during his presentations, for instance when presenting a photo showing an old iPod with a rotary dial as the new iPhone at minute 29:30 and waiting for the audience to react.

The third thing to note in this presentation is the way things are said. It is well known that Steve Jobs practiced his presentations many times until it looked effortless. Large numbers of backstage staff were involved, as well as a lot of technology, to ensure that the keynote address ran smoothly and without any major incidents. There are some rare examples of problems during Jobs' presentations and it is interesting to observe how he handled the problems. During this presentation, he speaks slowly, makes pauses, varies the speed at which he speaks, raises and lowers his voice, and repeats things to emphasize their importance. After 1:45:20 hours of the presentation, he wraps up by again displaying some of the most successful Apple products, followed by the new iPhone, repeats a quote from Wayne Gretzky, shows a historical photo of the company Apple Computers (now Apple Inc.) in its early days, thanks the audience, and concludes the presentation.

Garr Reynolds, who worked for Apple in the past as a Manager of their Worldwide User Group Relations, provides a second example of a good presentation. As a presenter he shows a lot more enthusiasm, makes many jokes, and communicates directly with his audience from the very first minute. A good example is the opening keynote lecture Reynolds gave at the 2009 Citrix Synergy conference in Las Vegas, organized by the Citrix Systems, Inc. software company:

```
http://citrix.com/tv/#videos/403
```

Let us analyze the first five minutes (approximately) of this 43:18 minute lecture. Of course, Reynolds' presentation is about presentations and not about a new product; he is teaching how to present, rather than introducing a new device. A science presentation is probably somewhere between these two extremes of teaching or introducing something new. "Hello Las Vegas", followed by a charming laugh and "How are you doing? Are you alright?", is the way he begins his presentation. He then continues by politely asking the audience whether they got enough sleep last night, and asks them to respond by raising their hands or their wine glasses. The first twelve seconds of the presentation aims to establish contact with the audience, to grab their attention, and to get people involved in the show, making very sure that they know that the presenter will continue to demand responses throughout the rest of the presentation. There is no escape from this highly interactive lecture!

It is strongly recommended that you watch the whole of the Reynolds lecture. Here, we look at a few examples of the presentation of the lecture, how the slides are designed, and how Reynolds stays in touch with his audience. Garr Reynolds believes in simplicity in slides, in the overall message, and in the way things are presented. He aims for high signal-to-noise ratios, as the example of the slide presented at minute 10:32 nicely demonstrates. Reynolds presents a very similar *xy* plot to the one that Steve Jobs used to describe smartphones in terms of how *easy-to-use* and *smart* they are, using it to explain that while software technology is growing more and more sophisticated it is also expected to remain simple to use. The *x*-axis represents the software sophistication and the *y*-axis the expectation of simplicity. A single straight line in this graph outlines the positive correlation between the two variables. While the graph remains on display, Reynolds explains the relationship several times, each time using different wording in order to make it quite clear to the audience.

The key message of simplicity in Reynolds' presentation appears after about 20 minutes when a slide is shown with a photo of a delicately balanced stack of rounded pebbles on the right and a quote on the left from the late artist, designer, and architect Koichi Kawana: "Simplicity means the achievement of maximum effect with minimum means". The slide itself is very simple, despite including text. The text is, however, given additional structure a few seconds after the slide is displayed by simply adding color to the second part of the sentence, "… maximum effect with minimum means". A very effective way to add structure to text in presentations, as well as on posters and in brochures, is to highlight keywords by making them bold, underlining them, or adding color. The presentation then provides numerous other examples of simplicity in slides, contrasted with examples of bad (complex) slides and, after minute 39:20, the example of Al Gore becoming a better presenter following training by Duarte Design Inc.

Nancy Duarte is the third example of a good presenter. She provides numerous video lectures and examples of presentation slides on her webpage, which also includes a blog. Her presentations are less sober than those given by Reynolds, more like the sort of presentations that you might expect from an artist or, if you like, more feminine. A good example of a video lecture can be found on her webpage under

```
http://duarte.com/speaking-engagements/
```

There is no problem at all with also using this more light hearted design of slides in a scientific presentation. I remember very well a presentation at the 1992 Annual Meeting of the Geological Society of America (GSA, http://geosociety.org), when Kenneth M. Schopf used the example of Mikado pick-up sticks to explain ecological locking and the stability of fossil morphologies (Schopf et al. 1992). During his presentation photos were shown in which sticks were picked up, provoking movement and reorganization of the remaining sticks, representing the changing interactions between organisms of the ecosystem. Whereas the scientific topic was very theoretical and difficult to understand, the presenter kept the audience attentive by showing great photos of Mikado sticks from time to time. Similarly, Axel Meyer, biologist at the University of Constance in Southern Germany, presenting at a symposium on the East African Lakes in Jinja, Uganda in 1993, broke up his talk on using the molecular-clock approach to dating the dispersal of East African cichlids by showing colorful fish on every third or fourth slide. More artistic slides using the hand-drawn effect of graphics or presentation software may have the same effect, making the presentation look more alive than a more sober presentation design. During her presentations Duarte also communicates with her audience, asks them questions, and maintains eye contact with individuals in the audience.

The way oral presentations are given very much depends on the discipline, the topic presented, and the audience attending the presentation, as well as reflecting the personal preference of the presenter (as in some of the examples provided). However, some fundamental rules that are independent of the discipline, topic, and audience apply to oral presentations and need to be kept in mind when planning such a presentation. The planning of a presentation includes the following three steps:

1. Deciding what message you want to deliver, and to whom. – Decide on the main points of your presentation, avoid unnecessary detail, and keep to a minimum the quantity of information to be delivered. The way the presentation is designed will depend on the individuals making up the anticipated audience, their levels of relevant knowledge, their interests, and their expectations.

2. Designing the concept and didactics of the presentation. – A good presentation has a clear and simple structure that will need to be defined before deciding on the contents of the slides. Paper cards can serve as analog slides that can be rearranged on a table when going through the concept of the presentation, prior to going digital. A template should then be created and a unified layout selected for all slides, including colors and fonts, before actually creating the individual slides.

3. Practicing and delivering the presentation. – Most people write down the text for a talk in order to better plan the presentation and to estimate its duration. If you do so, remember that memorized written text never works in conferences. A lot of practice is necessary instead, in order to become detached from the written text and to ensure a truly live presentation.

The following sections describe these three steps in greater detail.

9.3 Designing the Concept

As an example, let us design a two-minute presentation. At the University of Potsdam, where a course on collecting, processing, and presenting geoscientific information was held, the students organized one-hour sessions with 10–15 two-minute presentations chaired by the project leaders and their deputies. A large lecture hall provided a great conference-type atmosphere with its audio-video system, artificial-light dimming and window shading systems, and large projection screens. Two minutes may seem far too short a time in which to present anything, but experience from the course clearly shows that this is an ideal length for a talk by a beginner.

Assuming that we have simple slides with no text, three to five slides can be shown during such an oral presentation, depending on the complexity of any graphics included in the talk. Much has already been written on the best concept for a series of slides.

The classic idea for presentations, which has been stated numerous times elsewhere, is that you should "Tell them what you're going to tell them, tell them, and then tell them what you told them". Of course this is a good concept in general, but many talks have been rendered quite boring because presenters have adhered too strictly to this adage, and the talk that stands out in your memory is often the one in which the presenter dared to do things differently. We will come back to this point later when we see that a good presenter may indeed follow this golden rule, but in a way that is not always obvious to the audience. Boring examples of presentations, however, typically have the following structure:

1. The first slide shows a long title, many authors, the affiliations of the researchers, and the logos of any sponsors. While showing this slide, the presenter typically starts by thanking the co-authors and sponsors, and then explains the basic idea of the project and how it all started.

2. The next slide presents the outline of the talk, usually as a bullet point list, and although everybody knows what to expect the presenter tells the audience that he or she is going to present some background information on the project, the methods, the results, and an interpretation or discussion of the results, followed by a summary and an outlook for future research.

3. In the presentation that follows the presenter sticks to the outline and presents the results and interpretation, slide by slide. These include many text slides, bullet point lists, large tables, and complex graphs with small fonts and thin lines.

4. The most boring part comes at the end, when one (or even several) text slides are presented, giving a summary of the talk. In some rare cases the presenter, realizing that it would be very boring to read out all of the text, has been known to simply invite the audience to read it for themselves, rather than have the presenter read it out. In most cases, however, the

presenter really does read it all out, as if the presentation would not be complete without reading the summary.

You may, however, have noticed that none of the presenters discussed in the previous section followed this type of structure. Steve Jobs says things like "Every once in a while a revolutionary product comes along that changes everything" and then "Today, Apple is going to reinvent the phone". Later, he says that he will introduce an iPod: a phone and an internet device rolled together into a single unit. This indeed follows the rule to tell the audience what he is going to talk about. The difference, however, is that he does not explain the concept of the talk in the way that many people would do. Instead, he tries to attract attention and to increase anticipation by promising something exciting (a new device), before then going on to outline the new product, rather than just presenting an outline of the talk. Garr Reynolds does the same thing, spending a lot of time introducing the topic of the talk, rather than introducing the talk itself, after having first established contact with the audience through his opening remark.

In essence therefore, by all means tell your audience what you are going to talk about, but not how you are going to do it. Your colleagues certainly know already that you will show them your sample locations on a map, the equipment that you used to analyze the samples, graphical presentations of the results, and finally, your interpretation of the results.

To start a two-minute presentation on Neil Robert's surprising results of a dry Younger Dryas in East Africa (from Chapter 2) you could, for example, show as the first slide a beautiful photo of Lake Magadi and the text:

```
The Younger Dryas in Africa
Neil Roberts
University of Leicester

together with
Maurice Taieb, Philip Barker, Brahim Damnati, Michel Icole and
David Williamson
```

using the largest font for the title and a small font for the co-authors. If you prefer, you could also add a subtitle, such as

```
The Younger Dryas in Africa
Wetter, drier, or no change at all?
```

We criticized the title of the paper for not including any information on the way that the East African climate changed during the Younger Dryas. The title of the talk, however, could deliberately exclude this information in order to increase interest and to make the audience curious about the outcome of the investigations in the Magadi basin. Audience curiosity can be stimulated by conference presentation titles such as "Life on Mars", with people attending because they want to know if the title is a question or an answer, and whether the researchers have indeed found evidence for life on the red planet.

While the slide is displayed you could explain to an audience that is not familiar with the current discussion on the influence that the Younger Dryas had on tropical climates, that contradictory results had been obtained from previous investigations completed in various locations at lower latitudes. Nobody, however, had ever looked at paleoclimate records from East Africa, which was why Roberts went to Lake Magadi and collected a sediment core, so that he could look at the Younger Dryas interval in greater detail. The next slides would then follow a logical sequence, starting with a map showing the location of Lake Magadi and the sediment core that Roberts collected, then a photo of the core cut in half, the climate proxy record and age dates showing that the lake's water level was low during the Younger Dryas, suggesting a drier climate. If you wish to convince the audience of your story it is best to avoid any unresolved questions. On the other hand, if you expect to receive advice from the audience, consider your questions carefully. To wrap up you could say that, at least in East Africa, the Younger Dryas indeed seems to have caused an increase in aridity. After two minutes, you thank the audience and the chairperson asks for questions.

9.4   Creating a Template

Having prepared an analog draft of the presentation, as suggested by Garr Reynolds, we now launch the presentation software, such as *OpenOffice Impress*, *Microsoft PowerPoint*, or *Apple Keynote*. We first create a template for the slides in our presentation that defines their size, background, color schemes, animations, font types, and font sizes, as well as other layout elements such as logos, slide numbers and so forth (Fig. 9.3). All presentation software tools come with a gallery of design templates. Most of these, however, have been seen many times at conferences and should therefore not be used, except as an inspiration for our own template.

We then decide on the size of the slides, remembering that almost all presentations are in landscape orientation. Conference presentations are typically displayed on a screen using a digital projector. Projectors vary in both their resolution and their brightness. Common projector *resolutions* are Super Video Graphics Array (SVGA, 800 × 600 pixels), Extended

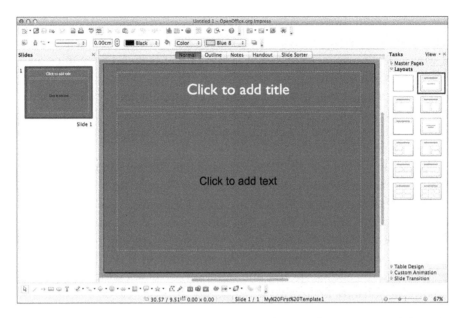

Fig. 9.3 Screenshot of the graphical user interface of *OpenOffice Impress*, including the *Slides* panel (on the left), the presentation slide itself (in the center), and the *Tasks* panel (on the right). This slide displays the template used for our presentation with a blue background and two text boxes.

Graphics Array (XGA, $1,024 \times 768$ pixels), and the two High Definition (HD) formats 720p ($1,280 \times 720$ pixels) and 1080p ($1,920 \times 1,080$ pixels). If you are not sure what the resolution of the projector to be used at the conference will be, and especially if you expect it to be an older device, choose an XGA resolution to be on the safe side. Only a few of the lower-cost or older devices that are used at workshops can handle only SVGA resolutions. At larger conferences, modern high-definition projectors are normally used rather than SVGA or XGA projectors.

Most modern projectors have a *brightness* of 1,500 to 2,500 ANSI lumen, which is suitable for small workshops or lectures, while more expensive projectors with a brightness of more than 4,000 ANSI lumen are used for large conference or lecture halls. The brightness of the projectors is an important aspect to consider when creating a template. Low brightness projectors in an insufficiently darkened room require a template with good contrast between background, text, and graphics.

Our example presentation will be created with the free *OpenOffice Impress* software, but working with Microsoft PowerPoint and Apple Keynote is very similar. The problem with OpenOffice, of course, is that it is not as stable as its commercial alternatives, it has no support line to call if you have a problem, and its documentation is not as complete as that of Microsoft PowerPoint and Apple Keynote. We launch OpenOffice and choose *Presentation* from the Start Center. The OpenOffice *Presentation Wizard* pops up and asks us to choose between an empty presentation, a presentation from a template, and an existing presentation. We will ignore the templates gallery and design our own simple template, and so we choose *Empty presentation.*

We then select a background for our template. Presentations are typically in landscape orientation although Microsoft PowerPoint also allows the use of portrait orientation, but not of mixed orientations. The Presentation Wizard displays a preview of the first slide and a list of background templates. We again ignore the templates and choose *<Original>* from the list and *Screen* as the output medium. The Presentation Wizard then allows us to choose the effects, speeds, and durations of slides, which we ignore. Finally, the software opens two windows side by side, one of which shows a presentation in the middle panel of a user interface with a toolbar, slide browser, and task center. The other window lists the styles and formatting of the slides.

We can now create a new template using *Save…* from *Templates* in the *File* menu and saving the new template as *My First Template.* We then select the *Slide Master* option from *Master* in the *View* menu. We choose *Page…* from the *Format* menu, and then select the *Background* tab. A pull-down menu lists *None, Color, Gradient, Hatching,* and *Bitmap,* from which we choose *Color* and then *Chart 1,* which is a classic background for presentations. Since the first release of Microsoft PowerPoint, color gradients from dark blue at the top to lighter blue at the bottom of the slide have been widely used by presenters, but these are now slowly going out of fashion. Alternatives are patterns, textures or (blurred) photos, but these need to be used with care in order to maintain a good contrast between the background and the content of the slide. Hatching is not normally used as a background for presentations. White backgrounds are, however, becoming increasingly popular as they avoid the need for frames around screenshots taken from published graphics with white backgrounds. Interestingly, some presenters use *Light green* or *Yellow* as their backgrounds, which are really painful to look at. To sum up the discussion on backgrounds, it is best to use either dark or very light backgrounds, in order to ensure good contrasts in the slides.

We can also use the *Slide Master* to change the default font. Always use sans-serif fonts such as *Helvetica, Arial* or *Gill Sans;* serif fonts contribute to the legibility and readability of printed documents such as newspapers and magazines, but are never used for online documents or presentations as the serifs often appear blurry on computer screens or displays. The master slide shows text fields with all defined styles such as *Title*, *Outline 1* to *9*, and so on. We can click in the *Title* field, mark the title, and then change the font type, size, and color on the master slide. As an example, we change all fonts to Gill Sans, leave the font size as it was but change the font color to white. You can also add logos and slide numbers, although it is strongly recommended that not too much information be included. We close the Slide Master and save the template as *My First Template* using *Save...* from the *Templates* menu within the *File* menu. Since *My First Template* already exists, the software asks us whether we wish to replace the existing template, and we click on *Yes*.

9.5 Creating Slides

We can now create a new presentation from the sample template. We again choose Presentation from the OpenOffice Start Center, choose *From template* from the Presentation Wizard, then *My Templates* from the pull-down menu, and finally select *My First Template*, which we have just created. The software creates a new document entitled *Untitled 1* that we can save as *myfirstpresentation_vs1.odp*. The first slide shows the same layout as before, with the *Chart 1* background and *Gill Sans* font.

We first create the title slide (Fig. 9.4). The first slide of an empty presentation has two text boxes. We delete the lower text box and move the upper one to the center of the slide. We can do this by right-clicking on the edge of the text box and centering the object both vertically and horizontally using *Alignment* from the pull-down menu. The title of our presentation is *The title of my presentation*. The author's name (*John Q. Scientist*) and his affiliation (*University of Nowhere, Department of Science*) appear in the next two lines below the title. We use the *Gill Sans 44 pt light white* font for the title and change the font for the author's details to *Gill Sans 26 pt light white* using *Character* from the *Format* menu. We then save the file as *myfirstpresentation_vs1.odp*.

The next slide is a text slide, although we will generally try to avoid text in our presentation. To start with, we again save the document but this time as *myfirstpresentation_vs2.odp*, thus creating version 2 of the document. We create a new slide by right-clicking in the *Slides* panel on the left. Alternatively, we can add a slide using *Slide* from the *Insert* menu. Instead of listing the outline of the talk as discussed in Section 9.3, we put a hy-

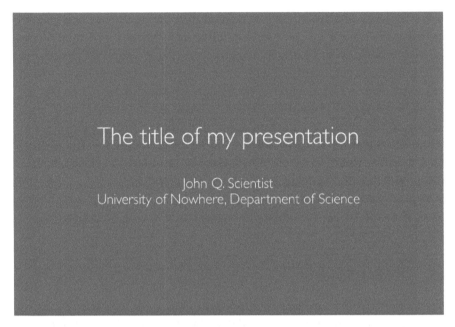

Fig. 9.4 The first slide of an imaginary presentation. The title of our presentation is *The title of my presentation*, the author is *John Q. Scientist* and his affiliation is with the *University of Nowhere, Department of Science*, as shown in the two lines below the title. We use the *Gill Sans 44 pt light white* font for the title and change the font for the author to *Gill Sans 26 pt light white*, using *Character* from the *Format* menu.

pothesis, a research question, or a scientific controversy on this slide. The *Layouts* panel on the right allows us to choose the *Centered Text* layout. Within the slide, the text box asks us to *Click to add text*. We type *What's the research topic?* into the text box and change the font to *Gill Sans 32 pt light white*.

The third slide contains a table, even though tables are not ideal for presentations. If a table is presented then it should be one with very few columns and rows, such as the one edited in Section 8.5. We again first need to save the document, this time as *myfirstpresentation_vs3.odp*, thus creating version 3 of the document. We then again add a new slide using *Slide* from the *Insert* menu and change the slide layout to *Blank Slide* in the *Slides* panel. The tab-delimited table in Section 8.5 has seven rows and three columns. We use *Table ...* from the *Insert* menu and choose the *Number of columns* as *3* and the *Number of rows* as *7*. After pressing *OK*, a table with a blue fill appears on the slide. Inserting the table from the clipboard (i.e. copied from the file *geochem_data.txt* used in Chapter 4), however, creates a separate text object instead of filling the new table fields with the elements from the original table.

We can of course type the numbers into the table fields manually and then edit the layout of the table. After typing in the numbers, we change the font to *Gill Sans 24 pt light black*. The *Table* toolbar that appears after inserting a table can be used to change the design of the table, or to insert and delete single rows and columns, and so on. Having pasted the tab-delimited table into a new text box, we can use *Character...* from the *Format* menu to define *Gill Sans 24 pt light white* as the font. Using *Ruler* from the *View* menu, we can view the rulers and then place left tabs at 4.5 cm and 9 cm. We can also change the font of the header to *regular* instead of *light*. An empty line can be added after the header, in order to create some space between the header and the table. By right-clicking on the edge of the text box we can center the table, both vertically and horizontally, using *Alignment* from the pull-down menu.

The next slide contains vector graphics, for example the pie chart or the *xyy* plot edited in Chapter 8. We again need to first save the document, this time as *myfirstpresentation_vs4.odp*, thus creating version 4 of the document. We again add a new slide using *Slide* from the *Insert* menu. In theory, there are two ways to include vector graphics in a presentation. We can either import them directly as vector graphics, or we can export them from our vector graphics editor as raster graphics and then import the raster graphics into the presentation software. Both methods have their pros and cons. For example, the raster version of a graphic image is essentially opaque and cannot therefore be overlain on other graphic images or text objects in the slide. On the other hand, a high-resolution raster version of a graphic object avoids the distortions in text and graphics that can be caused by software incompatibilities. This, however, is at the expenses of being less editable after inclusion in a slide.

As an example, we import the line graph of the ice core data from R.B. Alley (Section 3.5) that we plotted in Section 5.2 and edited in Section 8.2 (Fig. 9.5). In the previous chapter we created a version of the line graph with a white font for use in slides with dark backgrounds, saved as *icecore_lineplot_vs7_ai_fordarkbackground.eps*. Using *Open...* from the *File* menu, the plot is opened in *OpenOffice Draw* instead of Impress, but we can easily copy and paste the graph into our presentation. Holding down the *Shift* key, we can then proportionally scale the graph to fit the size of the slide. Finally, we can again center the plot by right-clicking on it and then using Alignment. We of course have to add a reference to the graph, e.g., *Data from Alley (2000)*. Using the *Text* button from the lower toolbar we can

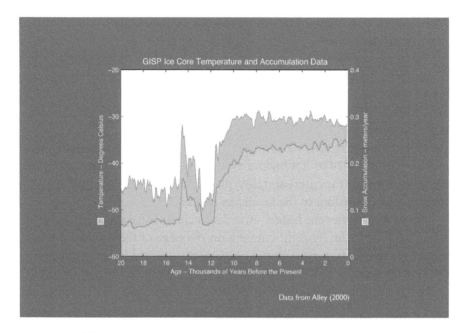

Fig. 9.5 The fifth slide in our presentation, showing the line graph of the ice core data from R.B. Alley (2000), using data from Section 3.5 that was plotted in Section 5.2 and edited in Section 8.2. We use a version of the line graph with white fonts, for use on slides with dark backgrounds.

place, edit, and move text objects within the slide. We can use *Character...* from the *Format* menu to define *Gill Sans 14 pt light black* as the font. We can also include, scale, and move a pie chart in our presentation, such as the one we created for use with dark backgrounds in the file *icecore_piechart_vs12_ai_fordarkbackground.eps*, in a similar way. In this case we do not need to add a reference since it uses our own data.

The very attractive surface plot, or block diagram, created in Section 6.3 and subsequently edited in Section 8.2 is obviously too complex to be integrated into an OpenOffice presentation (Fig. 9.6). It therefore provides us with a good example with which to demonstrate the conversion of a complex vector graphics file into a high-resolution raster image for inclusion in a presentation. The most common format used for presentation slides is the $1,024 \times 768$ pixel format (the Extended Graphics Array or XGA format - see Section 9.4 above). We can export the image as a raster image using any vector graphics software, and save it as a TIFF file. Since the original image created in Section 6.3 had dimensions of 10 cm × 10 cm, exporting it with a resolution of 300 dpi (dots per inch) gives us 1,181 pixels x 1,181 pixels. This

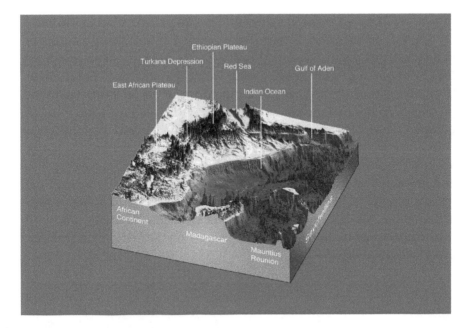

Fig. 9.6 The seventh slide in our presentation, with the very attractive surface plot, or block diagram, created in Section 6.3 and edited in Section 8.2. This image provides us with a good example with which to demonstrate the conversion of a complex vector graphics file into a high-resolution raster image, for inclusion in a presentation The most common format used for presentation slides is the 1,024 × 768 pixel format (the Extended Graphics Array or XGA format - see Section 9.4 above). We can export the image as a raster image using any vector graphics software, and save it as a TIFF file. Since the original image created in Section 6.3 had dimensions of 10 cm × 10 cm, exporting it with a resolution of 300 dpi (dots per inch) gives us 1,181 pixels × 1,181 pixels.

can then be saved as *etopo2_surfaceplotlight_vs7_ai_block.tif*. Again, using *Open...* from the *File* menu, the plot is opened in OpenOffice Draw and can be copied and pasted into our presentation.

The next slide contains raster graphics, such as a photo or a satellite image. We need to take care that we import the image with an appropriate resolution, in order to ensure that the image does not appear blurred or grainy in the presentation. We first need to save the file again, this time as *myfirstpresentation_vs5.odp*, thus creating version 5 of the document. We use the satellite image that was created in Section 7.4, edited in Section 8.3, and saved as JPEG file *naivasha_image_vs4_gimp_smallfile.jpg* (Fig. 9.7). We again create a new slide and open the image file in OpenOffice Draw. We can copy and paste the image from Draw into Impress, scale and center it, and then add a reference (*NASA/GSFC/METI/ERSDAC/JAROS and U.S./*

Japan ASTER Science Team) using *Gill Sans 14 pt light white* as the font.

The last slide comes with a take-home message in the form of a summary or an important photo. Never put conclusions as a long list of bullet-points. It is better use a graphic image and to tell people the conclusions, rather than putting them in writing and reading through the list at the end of the presentation. In our example we simply copy the slide with the research question, paste it after the slide with the satellite image, and change the text to our main conclusion. It is not a good idea to then add another slide with acknowledgements, showing a long list of the advisors, colleagues, friends, and sponsors that have all contributed to your work, in the same way that it is not good to include too many names and logos on the first slide of your presentation. You can briefly mention your colleagues and sponsors but it has become a bad tradition to put a lot of logos on the first slide, or even on every slide, of a presentation.

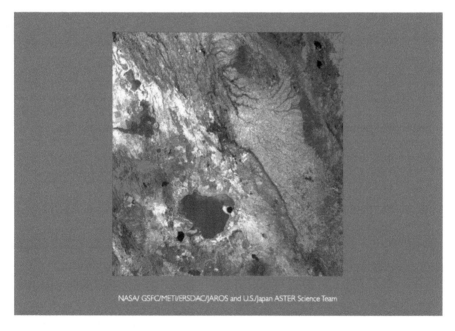

NASA/ GSFC/METI/ERSDAC/JAROS and U.S./Japan ASTER Science Team

Fig. 9.7 The eighth slide in our presentation, containing a satellite image. We need to take care that we import the image with an appropriate resolution in order to ensure that the image does not appear blurred or grainy in the presentation. We first need to save the file again, this time as *myfirstpresentation_vs5.odp*, thus creating version 5 of the document. We use the satellite image created in Section 7.4 that was edited in Section 8.3 and saved as JPEG file *naivasha_image_vs4_gimp_smallfile.jpg* (Fig. 9.7). We again create a new slide and open the image file in *OpenOffice Draw*. We can copy and paste the image from Draw into *Impress*, scale and center it, and then add a reference (*NASA/GSFC/METI/ERSDAC/ JAROS and U.S./Japan ASTER Science Team*) using *Gill Sans 14 pt light white* as the font.

Having finished creating our slides, we can now view them all using *Slide Show* from the *Slide Show* menu. Then, as must happen quite often at conferences, the presentation may not look exactly as we expected. I cannot predict how the presentation will look on your own computer, but some typical problems arise on mine when presenting at conferences. For example, all slides sometimes have blue backgrounds, even if I have changed some of them to white. The backgrounds may be white in the preview, but in presentation mode they are all blue. Another classic problem that arises at conferences when different computers are used for creating slides, and for presenting them, even if they both share the same operating system, is that vector graphics may only show as placeholder boxes, displaying the information that the graphics have been created with *Adobe Illustrator*, but no graphics. Even if you try to export the presentation as a PDF file using *Export as PDF...* from the *File* menu, the vector graphics will still not appear in the document. Once again, they are all nicely displayed in the preview mode of the software when editing the slides. The best way to avoid these problems is to export all vector graphics as raster images, merge the images into a single PDF file and then present the PDF document using the *Full Screen* mode of the PDF software, a strategy that has become very popular when there are presentation incompatibilities at conferences.

Now let us investigate the commercial alternatives to OpenOffice Impress, which are *Microsoft PowerPoint* and *Apple Keynote*. Both products offer tools that are much more professional than those provided by OpenOffice. Having been frustrated by OpenOffice Impress, we could simple export the entire presentation into *Microsoft PowerPoint* format under the name *myfirstpresentation_vs6.ppt*, using *Export...* from the *File* menu. Opening the file in PowerPoint reveals some minor changes or incompatibilities between the two software tools. For example, the *light* mode of the *Gill Sans* font has been lost, but we can easily change the font back to *Gill Sans light* within the software. Tables also cause the same problems as with Impress: text therefore needs to be converted into a table with *Microsoft Word* before being included in the presentation. We can remove the two empty boxes in the slides that should have contained the vector graphics. Including the vector graphics into the presentation is then very easy as they can simply be dragged from the file browser and dropped into the slide. Once they are in the presentation they can also be copied into other slides using copy and paste. Scaling and aligning the graphics objects is possible after double-clicking on them. Switching to the presentation mode reveals that this presentation works perfectly well, despite the inclusion of vector graphics.

As an alternative we can open the file *myfirstpresentation_vs6.ppt* with *Apple Keynote*. Drag and drop also works here for importing the vector graphics into the presentation. The positive surprise, however, is that copy and paste also works fine with the table when using Keynote. We simple create an empty table first, modify its layout, and then copy and paste the tab-delimited table from the *geochem_data.txt* text file created in Chapter 4 into the empty table in the slide. We can then change the text font to *Gill Sans 24 pt light black* and change the header text to *regular*. We can export the presentation as a *PDF* file in order to ensure that the document will be compatible with the computer in the conference room. If you wish to be absolutely sure that your presentation will run on the conference computer, you can also export the slides as raster images (*TIFF, JPEG*) and merge the individual images into a single *PDF* document using *Adobe Acrobat*, or any other *PDF* software.

It is, of course, very tempting to add animations to the presentation as most presentation software tools provide large catalogues of animations. Adding video or sound can also make presentations very attractive, for example a video recording of an experiment, or the sound of an ancient musical instrument. There have, however, been numerous bad examples or unfortunate accidents, such as the sound of applause between slides when an author accidentally selected the applause sound transition in PowerPoint but obviously never switched on the speakers of his own computer. The applause, however, was clearly audible after the presentation had been copied onto the laptop provided in the conference hall. Some software provides simple animations that can be very useful in particular instances, such as moving objects around, or changing their scaling. A good way to animate is to assemble an image in several stages, as in the previously discussed examples presented by Steve Jobs and Garr Reynolds, rather than presenting the complete graphic all at once. Another very useful feature allows the inclusion of links to internet webpages that provide access to additional resources, such as audio or video content.

9.6 Practice and Delivery

In Section 9.3 we designed the structure of a two-minute presentation. Two minutes is an ideal length for a talk by a beginner as it allows the talk to be practiced repeatedly. If talks are longer, beginners tend to practice the first five minutes quite well but neglect the rest of the talk.

Judith P. Rhodes in her booklet entitled *Scientifically Speaking* (which was first published in 1995 but then updated and revised by others in 2005) gives an excellent summary of, among other things, how to plan, design, and give an oral presentation. A beginner is advised to write down the entire talk from beginning to end, but reading this text at a conference is strictly forbidden. A presenter simply reading a text can be very boring for the audience. Giving an oral presentation is a method of communicating with your colleagues, and it is therefore important to maintain eye contact with the audience while speaking freely, rather than simply focusing on a sheet of paper.

Having written out your talk, you should read it aloud several times, ideally in front of friends or colleagues from your group. Reading aloud helps you to identify any parts of the text that sounds awkward. Moreover, reading the text right through several times helps you to measure the actual time that the talk will require. At conferences, it is regarded as rude and egotistical to run overtime. If the time allotted for your talk is 15 minutes including questions, you should plan your talk to last for only 12 minutes. However, be sure to take into account the effect of nervousness, which results in most people speaking more quickly during an actual conference presentation than when practicing.

After reading the presentation aloud several times you should start speaking freely, without referring to your written text. At this point it becomes very obvious if your talk has been well designed, and if the sequence of slides with graphics and photos tell the story nicely without the need for spoken words. In fact, the slides should help you to remember the things that you want to say. A good suggestion that is often made is to make a video recording of your presentation. Watching the video helps you check if you are talking to your laptop screen, rather than maintaining eye contact with the video camera (i.e. your audience). You can also check your body language, such as what you are doing with your hands, and how you are standing. According to Judith P. Rhodes, more than half of the interpersonal communication comes from facial expressions and body language, one third from vocal quality and the tone of voice, and less than ten percent from the content and the actual meaning of the words. In a science talk these relative proportions may well be different, but nonverbal signals clearly remain an important part of the communication process.

A two-minute presentation can easily be practiced 20 or 30 times before having a practice run in front of an expert audience. You should invite your audience to ask questions and to comment on the quality of the presentation. Once you are at the conference you should familiarize yourself with the room that you will use for your presentation. A final practice in that room would be ideal, but is not often possible; you can use a different speaker-ready room instead, if available.

The final delivery should be given using a friendly manner, voice, and body language. Never be aggressive or you will certainly lose your audience; even if they do not leave the room, they are sure to start reading their emails on their laptops using the conference center's Wi-Fi internet. While maintaining eye contact with your audience, try to smile from time to time! This often produces a great response, and individuals might express their agreement with your conclusions by nodding their heads or smiling back at you. By maintaining eye contact you may also notice if the audience has not understood your explanations, which you may then wish to repeat. In this way you will actually communicate with your audience, and your talk will be well delivered.

Recommended Reading

Duarte N (2009) slide:ology: The Art and Science of Creating Great Presentations. O'Reilly Media, Sebastopol, California

Duarte N (2010) Resonate: Present Visual Stories that Transform Audiences. Wiley, Hoboken, New Jersey

Reynolds G (2009) Presentation Zen Design: Simple Design Principles and Techniques to Enhance Your Presentations. New Riders Press, Upper Saddle River, New Jersey

Reynolds G (2010) The Naked Presenter: Delivering Powerful Presentations With or Without Slides (Voices That Matter). New Riders Press, Upper Saddle River, New Jersey

Reynolds G (2011) Presentation Zen: Simple Ideas on Presentation Design and Delivery. (Voices That Matter)–2nd Edition. New Riders Press, Upper Saddle River, New Jersey

Rhodes JP, Gargett A, Abbott M (2005) Scientifically Speaking. The Oceanography Society, http://tos.org/resources/publications/sci_speaking.html

Schopf K, Ivany LC, Morris PJ (1992) Onshore-Offshore trends in light of Ecological Locking. GSA Abstracts with programs, Cincinnati.

10 Creating Conference Posters

10.1 Introduction

The results of a project are typically presented in three formats; posters, talks, and papers. A poster is collection of figures, photos, and text printed on a large sheet of paper that is presented on a poster board in a large hall in a conference building. During poster sessions, the presenter of the poster can personally interact with the people attending the poster session and visiting the poster. This section is on planning and designing a poster, and also includes suggestions for practicing the presentation of a poster at a conference.

10.2 Planning a Poster

Careful planning is essential for a successful poster presentation. Poster sessions at conferences commonly include a large number of poorly designed posters, and thus a well prepared poster has a good chance of attracting attention. In her booklet entitled *Scientifically Speaking*, Judith P. Rhodes gives an excellent overview of how to plan, design, and present a poster. This booklet is highly recommended and is used as an inspiration for this section, together with further details added from our own experiences with posters at conferences.

The first objective of a poster is to attract people from a distance of 5 to 10 meters. Poster sessions are typically held in large halls in which conference participants are likely to be walking around, perhaps with a glass of wine, looking at posters in the late afternoon after long sessions of oral presentations. Within the poster areas, a large number of posters may be displayed on poster boards, with the authors standing by their posters, waiting for visitors and ready to discuss the results of their research. The presenters and their posters are always likely to attract people who are working in the same field and have found the poster in the abstract catalog, but the challenge is to also attract people working in other fields that are nevertheless related to the research topic. These are likely to be scientists that are attend-

M. H. Trauth and E. Sillmann, *MATLAB® and Design Recipes for Earth Sciences*, **223**
DOI: 10.1007/978-3-642-32544-1_10, © Springer-Verlag Berlin Heidelberg 2013

ing a poster session that covers a broad scientific field, such landslide risks in the Andes, life on Mars, or the causes and consequences of the Younger Dryas cold event.

There are many ways to attract attention to your poster, apart from your own personality, which of course always helps (large groups can often be seen hanging around posters presented by particularly attractive doctoral students). A well-designed poster, however, can also attract visitors even if your work is not exactly in their main field of interest. You then probably have just one or two seconds to stop a person moving on to the next poster. The two things most likely to help you attract people are a short, effective title and a good overall design. The title should be in a large font, delivering the main message quickly and completely (such as, for example, *There is life on Mars*, or *The oldest Homo ever found*), and a good overall design should incorporate nice colors, attractive graphics and photos, and not too much text.

Below the title you should list the authors, together with their institutional information (but without street addresses), in a smaller font than used for the title. Your email address, as the presenter of the work, should be included with the list of authors. It is a good idea to use first names in the list of authors as this will make it easier for people visiting your poster display to interact informally with you. Some presenters also display their photo on the poster to help visitors locate the author. In her booklet Judith Rhodes suggests omitting institutional logos, but a logo can be useful on a poster if it helps people to identify clusters of posters from the same institution as being related to each other within a larger project. Not many logos should be included, however, and of course not the logos of all of your sponsors,

Having attracted somebody's interest through your great title and design, you need to be able to quickly deliver your main findings and conclusions. If you are at your poster you will probably give a brief oral presentation of your work. In many cases, however, people will spend only two or three minutes at your poster if they are particularly interested, and probably less if your work is only peripheral to their interests. Guiding your visitor through your poster by numbering the figures or using arrows to indicate the sequence helps to keep their attention and encourages them to read the full story. The figure captions should of course always be kept very short. The most common mistake with posters is to include too much information, too many figures, and too much text. Reproducing the original abstract on a poster when it is already included in the conference's abstract volume (or CD) is pointless as most people will not bother to read large amounts of text on a poster anyway. A very brief introduction to the topic, followed by a series of figures with short captions and a brief concluding paragraph, is the maximum amount of text that should be included on a poster. The story should be told with graphics and photos, rather than with text. When assembling the poster, graphics and text should be large enough to be read from 2–3 meters away. Busy backgrounds such as complex photos should be avoided, and dark colors should be used on light backgrounds rather than the other way round.

The next two sections provide guidelines for creating a poster template from scratch, using *Inkscape* or *Adobe Illustrator* software. We then incorporate graphics and photos edited in Section 8 into the poster. The final section in this chapter then deals with presenting a poster at a conference.

10.3 Creating a Poster Template

A poster generally comprises a background (either white, a single color, or a background photo), a title and list of authors (including their institutional information), an abstract and/or a brief introduction to the topic, a selection of figures and photos with captions, a brief concluding paragraph, and finally, references and acknowledgments, if appropriate. In this section we decide on the background, select the colors to be used throughout the poster, and choose the sans-serif font to be used (e.g., Helvetica, Calibri or Arial). We will use Adobe *Illustrator* in this demonstration as it is the most professional software for the demanding task of creating a poster. If Illustrator is not available, however, we can also carry out all the same steps using the *Inkscape* open source software. Using an extreme landscape format we will present the title, subtitle, author and contact information, a short abstract, references, a table, and five images with captions, but no additional text. We will use a four column layout, boxes with rounded corners, a blue color scheme, and a tinted background. More general information on the software, including a User's Guide, detailed help, tutorials, and an explanation of technical terms, is available from:

```
http://help.adobe.com/en_US/illustrator/cs/using/index.html
```

Before using the computer we need to draft an outline of the poster on a piece of paper, which can even be as large as the actual poster, in order to plan its design and content. This outline should contain drafts of figures, images, and text, so that the visual structure of the poster can be checked. The design of the poster and the material presented can help to attract browsing visitors, who may not be particularly interested in your topic but could still contribute to your work through discussions from an outsider's perspective.

These can sometimes even be more inspiring than the comments that you receive from experts, which may well be comments that you are already familiar with.

A visit to a poster by a conference participant usually follows a more or less typical pattern. Knowing this pattern can be very helpful when creating the poster template. A visitor strolling through the poster session will decide in less than five seconds whether or not a poster looks interesting, so we need to design a template that can capture attention within this short period of time. The poster therefore needs to produce an appealing and cohesive first impression from a distance, which requires a clear structure, great figures, and an esthetic color scheme.

Approaching closer to the poster, a concise title that is intriguing, or even provocative, easy to read (in a large font size), will certainly arouse the curiosity of the visitor. Having first been attracted by the visual effect and the title of the poster, the visitor is likely to be interested in the authors and their affiliations, and then to take some time to actually look at individual graphics and photos. Most people do not want to read very much text and instead start looking around for the author of the poster to talk to. This would be the ideal result, with your poster putting you in touch with a visitor that you might not otherwise have met.

Following these more general considerations, we will now create the template for our poster. We first need to know the size of the conference's poster boards. This information is typically provided in the *Author Guidelines* on the conference webpage. As an example, we can read in the Author Guidelines for the *General Assembly 2010* of the *European Geoscience Union* (EGU) at

```
http://meetings.copernicus.org/egu2010/guidelines/author_
guidelines_poster.html
```

the following information:

```
Poster Boards
Poster boards are in landscape format with a clear dimension
of 197 cm width x 100 cm height. All the material necessary for
attaching the poster to the poster board is available at the
Facility Desks in the respective poster area. In addition, there
are assistants to help authors in putting up or in taking down
their posters. For each poster board there is a desk for placing
the private notebook/MacBook for additional PowerPoint or video
presentations. European-type power sockets are available.
```

The maximum size of the poster is therefore 197 cm × 100 cm. Depending on the size of your printer, the actual size of your poster may well be smaller, but must never be larger, than this maximum size. We now launch Illustrator and create a new document named *myfirstposter*, with a *Width* of 1,970 mm and a *Height* of 1,000 mm, with the *bleed* set to 0 mm. We choose the CMYK color mode because we will later want to print it out using an inkjet printer. Additional information on color modes and profiles is provided in Chapter 11. For the background we draw a rectangle with blue fill, creating a new color with the CMYK code C = 35, M = 20, Y = 10, K = 20 by using the swatches panel and the corresponding tool from the tool bar. For the name and logo of your institution or university, we create a rectangular banner on the left side of the poster, 120 mm wide and the same height as the poster, using a blue fill with C = 90 M = 50 Y = 20 K = 30. In order to achieve a uniform arrangement of objects, we choose *View > Show Grid*, open the preferences dialog box, and choose *Edit > Preferences > Guides & Grid* (in Windows) or *Illustrator > Preferences > Guides & Grid* (in Mac OS) and set the spacing between gridlines using *Gridline Every* 70 mm and *Subdivisions* 14.

Within the blue background we create two boxes in the left half of the poster for figures and text, each measuring 420 mm × 720 mm, and two more boxes in the right half, each measuring 420 mm × 920 mm. We fill all boxes with the blue color C = 10, M = 5, Y = 5 K =5, which is slightly paler than the blue of the background. We then select all boxes and use *Effect > Stylize > Round Corners* to round the corners of the boxes using a *Radius* of 15 mm, and define a white outline with a width of 20 pt. Having created the boxes we then arrange them provisionally (Fig. 10.1).

We need to leave enough space above the two boxes to the left for the title, subtitle, and author information, which we now add. We will then create an auxiliary construction of two rectangles and a system of four *guides* in a new layer we name *guides*. We then create two rectangles in a new layer for which we use a bright color such as M = 100, with one rectangle measuring 155 × 45 mm in the lower left corner of the poster and the other measuring 35 × 35 mm in the upper right corner. To create guides flanking the rectangles, we position the pointer on the left ruler for a vertical guide and drag the guide into position. Then we position the pointer on the top ruler for a horizontal guide and drag the guide into position. As a result, both rectangles are flanked by two horizontal and vertical guides each. Using *View > Guides > Lock Guides, Hide Guides,* or *Clear Guides*, we can edit these guides. The guides help to align the boxes vertically and horizontally, starting with moving the extreme left and the extreme right boxes to their correct positions, and then aligning the other boxes between the outer boxes using the *Horizontal Distribute Space* and the *Bottom Alignment*

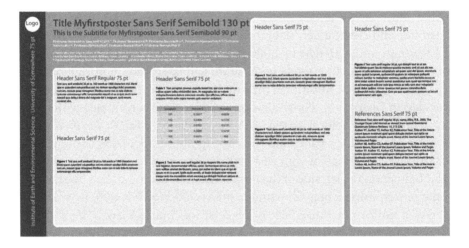

Fig. 10.1 Screenshot of the poster showing poster layout with a provisional distribution of text elements and boxes on a poster size of 197 cm × 100 cm.

buttons in the *Align* panel. The *Stroke* panel then is used to change the box outlines to white, with a 20 pt *Weight*. Having created and organized various text elements, the boxes, the sidebar, and the background color in the *Layers* panel, we delete the layer *guides* because the auxiliary construction is no longer required. Finally we save the document as the poster template *myfirstposter_1970_1000_empty.ait* and close the document.

10.4   Final Assembly of the Poster

Having created the poster template, we now assemble the poster to include the following figures, photos, and placeholder text created in previous sections:

`naivasha_image_vs7_ps_cmyk_2500px.jpg`	*1st column, Figure 1*
`icecore_piechart_vs11_ai_mask.eps`	*2nd column, Figure 2*
`icecore_lineplot_vs5_ai_cmyk.eps`	*3rd column, Figure 3*
`etopo2_filledcontourplot_vs1_matlab.eps`	*3rd column, Figure 4*
`etopo2_surfaceplotlight_vs10_ai_flattened.pdf`	*4th column, Figure 5*
`geochem_data.txt`	*2nd column, table*
`myfirstposter_placeholdertext.rtf`	

Opening the template again creates a new document, which we save as a new file called *myfirstposter_1970_1000_vs1_ai.pdf*. In the layer called *text* we create a new *text frame* and use *Copy and Paste* or *File > Place* to import the placeholder text onto the *canvas* surrounding the *artboard*. The *canvas* is a space where we can prepare elements like text and artwork, before moving them into the *artboard*, which is the only area that will be printed later. The plus sign next to the lower right corner of the *text frame* indicates overflowing text. We can click on it to create additional text frames linked to the first frame, providing sufficient space for the overflowing text. We change the style of the text (e.g., font type and font size) to improve its readability and check carefully to ensure that all glyphs or special characters have been correctly imported. As stated previously, we use the same sans serif font for all text on the poster (e.g., *Helvetica, Calibri, Myriad,* or *Arial*). Helvetica is a very popular 50-year-old Swiss font that is pre-installed on all Apple Macintosh computers, whereas PCs with Microsoft Windows installed provide the Arial alternative developed by Microsoft. The Myriad Pro font comes with the Adobe Creative Suite. Calibri is provided by MS Office and can also be freely downloaded from the Microsoft website.

We then create another new text frame in the text layer, and copy and paste the first line of the placeholder text *Institute of …* into it. All placeholder text includes information on a possible font size (75 pt in our example), which we can use by typing the numerical value of 75 pt into the *Character* panel. Using Object > Transform > Rotate we rotate the text frame, including the text, by 90°. Activating the *Preview* checkbox in each panel dialog window allows the results of all actions to be viewed prior to their actual application. If the text frame is too small to display the text, we can easily increase the size of a text object using the *Selection* tool and drag a handle on the bounding box. To make the bounding box visible, we choose View > Show Bounding Box. We then choose a white color for the text and move it into the dark blue sidebar, aligned with the lower horizontal guide and the vertical center of the blue rectangle of the banner. Above the text we place the logo of our university or institution, which is preferably a white vector graphics object. If an isolated vector graphics object is not available, we can place a raster image of the logo over a white background square. In this case, the logo resolution will need to be sufficiently high, e.g., 150 dpi.

We next insert the text *Title Myfirstposter … University, Town, Country* into a new text frame, again using a dark blue color for the *Title* and *Subtitle* and the same font sizes as indicated in the placeholder text. The author information should again be in a white font. Footnotes marked by characters such as *a*, *b*, … or an *asterix* can be used to link the authors to the relevant author information. We then create more text frames for *Header Sans Serif Regular 75 pt*, for the figure captions (such as *Figure 1 Text …*), for the table captions, and for the list of cited references, to be placed in the various boxes on the poster.

CREATING CONFERENCE POSTERS

10

Using *Window* > *Type* > *Paragraph Styles* and *Character Styles*, we open two more panels, which allows us to reuse the paragraph and character styles that have been previously defined, for a faster and consistent workflow. After selecting and formatting a text, such as a header, we click the *New Style* button in the *Paragraph Styles* panel, which opens the *Paragraph Styles Options* in which we can modify all settings, thus defining a style. After defining all styles for headers, the main text, references, and a bold *Character* style, these can then be used by selecting a text fragment and clicking on the desired style name in the associated panel. For the references we choose indents of 50 pt for the *Left Indent* and –50 pt for the *First Line Indent*. We then save the document as *myfirstposter_1970_1000_vs1_ai.pdf*, activating the *Preserve Illustrator Editing Capabilities* option and, if we wish, save a second version of the document as a poster template.

We next insert a table into the second box. Unfortunately, there is no simple way to format a table in Illustrator, which means that all tables need to be edited using other software, such as Microsoft Word or Apple Pages, prior to their inclusion in a poster. The table content is from the *geochem_data.txt* text file created in Chapter 4, which has already been used elsewhere in this book. We can place the edited version of this table, *geochem_data_table.pdf*, which has white separation lines, into the second box in the text layer. For the table background we choose a layer below the table, draw a rectangle using the background color for the table header, and a second rectangle using the new CMYK color (C = 20, M = 10, Y = 5, K = 10) for the table body. We then save the document as *myfirstposter_1970_1000_vs2_ai.pdf*.

We now insert the graphics and photos into the poster. To do this we create a new layer which we call *figures*, and use *File* > *Place* to place all figures listed above into this layer, and then move the figures to their desired positions and adjust their sizes. The figure *etopo2_filledcontourplot_vs1_matlab.eps* has not been modified since it was created with MATLAB. It looks like the original shoreline in the *coastline_linegraph_vs1_matlab_17703vertices.eps* file but has a much smaller number of vertices and a relatively small file size (1.3 MB).

MATLAB tends to split long paths when writing postscript files in order to save memory, and has done so in this file. In contrast, generating filled outlines (e.g. when using the `contourf` function in MATLAB see Section 6.3), creates closed polygons, which we position in a lower layer. The large number of vertices (17,703 in our example) may cause problems when trying to print the complex polygons using a PostScript printer, and also when including the figure in presentation slides. As described in Section 6.2, we need to reduce the vertices in order to overcome this problem, for instance, by using the MATLAB script presented in that section. In Illustrator the vertices are called *nodes* and their numbers can also be reduced in that software. As mentioned in Chapter 8, *Inkscape* seems to be unable to open complex vector files, producing the error message *Failed to load the requested file*. As an alternative we can export the critical zones as bitmap files and combine the bitmaps with a vector coordinate system, as shown in *etopo2_filledcontourplot_vs2_ai.pdf* and ... *vs3_ai.pdf*. After adjusting the size and the position of the figures we save the poster as *myfirstposter_1970_1000_vs3_ai.pdf*.

As a further modification of the poster in order to achieve a uniform appearance, we first draw another rectangle, this time 30 mm high, to be used as an auxiliary construction, as above. Using the pointer, this rectangle can be dragged across the workspace like a ruler to measure a constant horizontal distance of 30 mm between headers and figures, and between the white borders of the boxes and the text objects. Wherever we need to, we can drag

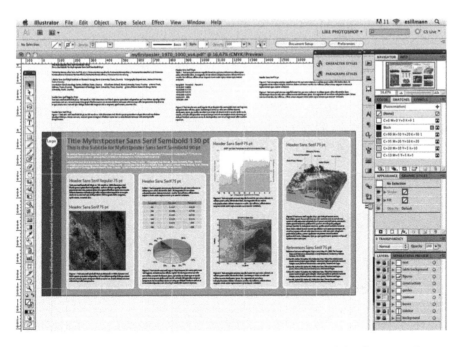

Fig. 10.2 Screenshot illustrating the assembly of a poster using Adobe Illustrator, showing the program's *menu bar*, several *panels*, horizontal and vertical *rulers*, *guides*, and the preparation of text elements on the *canvas*.

a guide (Fig. 10.2) from the horizontal *ruler* to help us maintain regular distances on the poster. We then save the poster as *myfirstposter_1970_1000_vs5_ai.pdf*, choosing a suitably *high PDF quality* preset, (such as *Press Quality* or *PDF/X-1a (2001 and 2003)*, or custom PDF preset files received from our service provider with the file extension *.joboption*), depending on the printer to be used. If the printer prefers to print outlines rather than text, we select all text and choose *Type > Create Outlines*. This is a useful way to avoid printer problems with special fonts, formatting, or glyphs that might not be embedded correctly into a PDF. In this case we save the poster file as version 6, making sure that we also retain the editable version 5 for possible later modifications to the text.

We may also want to create a compact PDF with a reduced file size for digital distribution, and therefore save version 7 in RGB color mode, choosing the *Smallest File Size* option from Adobe PDF preset. This allows us to send this file by email, to enable its contents to be discussed with a supervisor or with colleagues. It also enables us to test the visual effect by asking people from other disciplines for their feedback.

Before printing the final, full sized version of the poster (Fig. 10.3), we print some smaller versions to check the contents and layout. We can first print a low-cost, simple version for proofreading, using a laser printer. A second version can then be printed using the actual printer and paper that we will use for the final version but in a smaller size and using less ink,

Fig. 10.3 Final version of the poster, with background, sidebar, and boxes in blue shades, placeholder text of ~ 4600 letters including blank spaces, five illustrations, and a table. The graphics were generated in previous chapters and edited in Chapter 8.

or we can just print out part of the poster by defining an appropriate section using the *Artboard Tool* in Illustrator. Many different types of paper are available for printing posters with an inkjet printer. The standard office paper normally used with desktop printers is of uncoated, matte quality and has a weight of 21 to 27 lb (80–100 g). Posters are usually printed on slightly heavier paper with a weight of around 55 lb (200 g), either with an uncoated and matte surface or with a slightly more expensive coated, glossy or semi-matte, surface. For the highest quality printout, coated paper generally shows a brighter surface, a reduced absorbency of ink, and brilliant colors, similar to the inkjet paper used for printing photos. A coated, but semi-matte paper reduces reflections and is therefore best suited for posters that include text, figures, and photos, to be presented in a large foyer or hall with complex light sources.

10.5   Presenting a Poster at a Conference

Posters are presented in large foyers or halls, often with large numbers of other posters, organized according to themes that usually relate to oral sessions presented on the same day. The participants in the oral sessions, including the conveners, the presenters, and the audience, can visit the posters, meet with the authors, and discuss with them their results. The conference organizers often offer free wine and soft drinks, making the poster sessions an attractive opportunity to meet other scientists and discuss the overall topic of the poster session. These poster sessions are also important recruiting events and provide many opportunities for young researchers, such as doctoral students seeking postdoctoral positions for after the completion of their theses. Since the time slots available for oral presentations at large conferences are often very limited, most abstracts submitted with an oral preference actually end up as posters instead.

The models used to organize a poster session can vary from conference to conference, and even within a single conference, depending on the convener. Some conveners allow poster presenters to give a one-minute presentation of their work before the relevant oral sessions, possibly even including the presentation of a single slide. Others organize guided tours through the poster session, usually consisting of a small group of interested scientists going from poster to poster, with the presenters giving a short oral summary of their work and taking questions from the group. In most cases, however, the conference participants wander around individually and discuss the science with the authors at their posters.

How should one prepare for a poster presentation? Of course presenting a poster to an interested individual or a small group of people is a much less challenging task than giving a talk in a large lecture hall. Nevertheless, an inexperienced presenter should write down a few essential facts and conclusions from the poster, and prepare a quick summary to be presented when an interested visitor shows up at the poster. A very popular idea is to offer small printouts of the poster (e.g., using A4 format) that people can take home or, if available, reprints of a published journal article that contains the results shown on the poster or related work by the presenter. Some people also like to offer their business cards at their poster boards, or even their CVs if they are looking for a job. Laptops are sometimes placed next to the poster board to display material such as movies, computer animations, project webpages, or software tools.

During the actual poster session, which will typically last one or two hours, the authors of the posters are expected to remain at their poster boards, making it difficult for them to visit other posters in the same session. If there are two or more authors presenting the poster they can obviously take turns at the poster board. If only one author is present, he/she can put a note on the poster informing visitors when they will be available to answer questions.

Recommended Reading

Rhodes, JP, Gargett, A, Abbott, M (2005) Scientifically Speaking. The Oceanography Society, http://tos.org/resources/publications/sci_speaking.html

11 Creating Manuscripts, Flyers, and Books

11.1 Introduction

Oral and poster presentations are typically held at conferences or workshops, with a limited number of attendees. Making posters and presentations available online, as either PDF files or videos, increases the potential size of the audience. The most effective way to present scientific information is, however, to publish it as a journal article, thesis, or book. In contrast to these multipage formats, flyers and brochures are used to provide scientific information in a highly condensed form and are typically distributed as handouts at conferences, from trade show booths, during roadshows, or at events presented for the general public.

This chapter deals with the creation of manuscripts, flyers, and brochures as examples of published text documents. We first discuss how to plan and compose a manuscript, and then demonstrate how to create the ready-to-print document. For the first part we use a published research article by John Westgate from the University of Toronto, as a textbook-type example of how to structure a manuscript. The second, more technical part introduces the use of desktop publishing tools to lay out the manuscript.

11.2 Planning a Manuscript

Having completed the experimental part of a project, and then presented and discussed the results at conferences or workshops, scientists are generally required to publish the results of their work as journal articles, theses, project reports and proceedings, or in books. In most cases the manuscripts need to follow certain guidelines, which are usually summarized in a *Guide for Authors* provided on the publisher's webpage. The Guide for Authors for Elsevier's *Quaternary Research* journal, for example, can be viewed online at

```
http://www.elsevier.com/wps/find/journaldescription.cws_
home/622937/authorinstructions
```

M. H. Trauth and E. Sillmann, *MATLAB® and Design Recipes for Earth Sciences*, **235**
DOI: 10.1007/978-3-642-32544-1_11, © Springer-Verlag Berlin Heidelberg 2013

This guide provides all the important information that an author requires on the type of paper to be used, how to prepare and layout the content, and where to submit the manuscript. The author's guidelines first explain the type of work typically published by this journal:

```
Quaternary Research is devoted to interdisciplinary articles
dealing with the Quaternary Period. Articles must be of broad
interest, be of basic significance to more than one discipline,
and constitute a significant contribution to knowledge. Suitable
contributions include previously unpublished research results
and comprehensive reviews or syntheses of a field of knowledge.
(cont'd)
```

This information, which is available for any journal, helps authors to choose an appropriate journal in which to publish their work. Other criteria such as impact factors are also of great importance in choosing the best possible journal (see Chapter 2). In the next paragraph, the guidelines advise authors that the editor does not accept manuscripts exceeding 6,000 words. Multidisciplinary journals such as *Nature* and *Science* only publish much shorter manuscripts. *Science*, for example, only accepts manuscripts of up to 4,500 words (including 40 references) as Research Articles, or of up to 2,500 words as Reports. In most cases the Guide for Authors also contains information on the required structure of a manuscript, which for most journals is as follows:

```
Title
Author names and affiliations
Abstract
Introduction
Regional Setting
Materials and Methods
Results
Discussion
Conclusion
References
```

A research article includes tables and figures with captions, acknowledgements (including information on funding, research permits, and colleagues that have contributed to discussions of the work), appendices, and additional supplementary material.

In Chapter 2 we analyzed an article by Neil Roberts published in *Nature* in 1993. We saw how the structure of an article, including its title, the abstract (consisting of the first paragraph), the figures, and the conclusion help the reader to quickly grasp the most important aspects of the research. Since *Nature* and *Science* articles use a slightly different format from that explained above, let us use the article

```
Westgate JA, Shane PAR, Pearce NJG, Perkins WT, Korisettar R,
Chesner CA, Williams MAJ, Acharyya SK, 1998, All Toba tephra
occurrences across peninsular India belong to the 75,000 yr BP
eruption. Quaternary Research, 50, 107-112.
```

published in 1998 on pages 107–112 in volume 50 of the Elsevier *Quaternary Research* journal. This paper has been chosen to explain the structure of a research manuscript because it is a classic example of a journal article with a single conclusion, which is effectively summarized in the 14-word title:

```
All Toba tephra occurrences across peninsular India belong to the
75,000 yr BP eruption.
```

Most readers need two or three seconds to read and understand the title. Indeed, there is not much to add if you are only interested in the age of the Toba tephra that occur in India. The title, which is very concise, includes the location (*India*), the material (*Toba tephra*) and the age (*all ... belong to the 75,000 yr BP eruption*).

The abstract of the article contains additional information on the motivation for the work, the design of the research, the methods employed, the results, the conclusions, and the significance.

```
A controversy currently exists regarding the number of Toba
eruptive events represented in the tephra occurrences across
peninsular India. Some claim the presence of a single bed,
the 75,000-yr-old Toba tephra; others argue that dating and
archaeological evidence suggest the presence of earlier Toba
tephra. Resolution of this issue was sought through detailed
geochemical analyses of a comprehensive suite of samples,
allowing comparison of the Indian samples to those from the
Toba caldera in northern Sumatra, Malaysia, and, importantly,
the sedimentary core at ODP Site 758 in the Indian Ocean - a
core that contains several of the earlier Toba tephra beds. In
addition, two samples of Toba tephra from western India were
dated by the fission-track method. The results unequivocally
demonstrate that all the presently known Toba tephra occurrences
in peninsular India belong to the 75,000 yr B.P. Toba eruption.
Hence, this tephra bed can be used as an effective tool in the
correlation and dating of late Quaternary sedimentary sequences
across India and it can no longer be used in support of a middle
Pleistocene age for associated Acheulian artifacts.
```

The structure of the 184-word abstract is as follows. The first sentence, starting with *A controversy currently exists ...*, presents an introduction to the topic and explains why is it important to take the time and effort to analyze the material. The second sentence, which starts with *Some claim ...*, outlines the current hypotheses to be tested. The next two sentences, from *Resolution of this issue ...* to *... were dated by the fission-track method*, provide information on the nature of the research undertaken and the analytical techniques employed to perform the research. The sentence *The results unequivocally demonstrate …* then introduces and discusses the results. Finally, the conclusions and implications of the results are presented in the last sentence: *Hence, the tephra … can no longer be used in support of a middle Pleistocene age for associated Acheulian artifacts.* From this analysis the abstract can be seen to be a very short version of the full research article, structured as follows:

```
Introduction (including location)
Materials and methods
Results
Discussion
Conclusion
```

The third level of detail on the research results is then provided by the main text. The *Introduction* is again structured in a way that allows the reader to quickly access the required information. The first few sentences provide an introduction to the topic and some historical background on previous work. The sentence *The number of Toba eruptive events … is presently debated.* introduces the scientific problem to be tackled by the research, followed by a summary of the current hypotheses and the significance of the project. The Introduction closes with the sentence *Here, we attempt to answer the question …*, but most articles use *Here, we present …* instead, followed by a description of the research, presentation of the results, a discussion, and the conclusions.

In field-based research articles, the Introduction is typically followed by sections describing the geographic and geologic settings of the study area, including its topography, rivers and lakes, climate, vegetation, lithological units, and tectonic structures. In the paper by Westgate (1998), the information on the setting is restricted to a location map and a table listing all samples that were analyzed during the project, since the detailed setting is not relevant in this instance. Instead, the sections entitled *Composition of Glass Shards* and *Age Data* contain information on the analytical methods employed, which included the use of an electron microprobe, ICP-MS analyses, and fission-track dating. In this very short paper, the results of the chemical analyses and age determinations are also presented within these sections, including three tables and a ternary diagram. The *Age Data* section includes some discussion of the results, concluding that all Toba tephra are from the same eruption. The *Conclusion* summarizes the results with the statement that *New … data show that all the Toba tephra occurrences … belong to the 75,000-yr-old YTT bed.* and provides information on the implications and significance of the work. The article then closes with acknowledgements and a list of the references cited.

This analysis of the article by Westgate (1998) should help you to plan your own research article, using a similar structure. Writing a manuscript involves first preparing an outline, often comprising a series of figures that tell a simple story based on a new data set, the results of new research, or a review of existing data. A first draft usually has a maximum length of one or two pages and is structured in the same way as the full manuscript. The draft needs to be discussed with the co-authors and other collaborators in the project before starting to write more detailed text. It is recommend that the first author then creates the full version of the text, for example as *myfirstpaper_vs1.odt*, with OpenOffice Writer or *myfirstpaper_vs1.doc* with Microsoft Word, sends the file to the co-authors who make any changes that they require and then add their initials to the file name (e.g. *myfirstpaper_vs1_mht.odt* or *myfirstpaper_vs1_mht.doc*) before returning the file to the first author. Once all co-authors have agreed on the final version of the manuscript, the first author submits it to the chosen journal.

When writing the manuscript with a text processor such as OpenOffice Writer, Microsoft Word, or Apple Pages, it is strongly recommended that the character and paragraph formatting feature of the software be used. To demonstrate the use of character and paragraph formatting, let us assume that the manuscript has a classic structure and is composed of the following sections:

```
Title
Author names and affiliations
Abstract
Introduction
Regional Setting
Materials and Methods
Results
Discussion
Conclusion
References
Figure Captions
```

Since we have not yet written a scientific text, we use a placeholder text instead. The most popular placeholder text in desktop publishing examples is the free pseudo-latin text *lorem ipsum* (see http://en.wikipedia.org/wiki/Lorem_ipsum for details on the history and discovery of this text):

```
Lorem ipsum: Opera sine nomine scripta

Lorem ipsum dolor sit amet, consectetur adipisici elit, sed
eiusmod tempor incidunt ut labore et dolore magna aliqua. Ut enim
```

```
ad minim veniam, quis nostrud exercitation ullamco laboris nisi
ut aliquid ex ea commodi consequat. Quis aute iure reprehenderit
in voluptate velit esse cillum dolore eu fugiat nulla pariatur.
Excepteur sint obcaecat cupiditat non proident, sunt in culpa qui
officia deserunt mollit anim id est laborum.
```

The authors' guidelines for particular journals provide all the necessary information on formatting the manuscript file, but are often not very detailed with regard to the formatting of the text. For example, the Guide for Authors for the *Quaternary Research* journal states that

```
Manuscripts should be double-spaced and left-justified throughout.
Pages should be numbered consecutively and include line numbers.
```

without providing any further information on the font and font size to be used. We launch *OpenOffice* and choose *Text Document* from the Start Center to start *OpenOffice Writer*. The OpenOffice Writer displays a window with a toolbar and a blank page called *Untitled 1*, with text boundaries and a blinking cursor in the upper left corner, within the text box. As an example, we import the following piece of text from the *myfirstpaper_vs1.txt* file using *Open ...* from the *File* menu:

```
This is the title of our manuscript
John Q. Scientist, Jane W. Researcher
University of Nowhere, Department of Science
Abstract
Lorem ipsum dolor sit amet, consectetur adipisici elit, sed
eiusmod tempor incidunt ut labore et dolore magna aliqua. Ut enim
ad minim veniam, quis nostrud exercitation ullamco laboris nisi
ut aliquid ex ea commodi consequat. Quis aute iure reprehenderit
in voluptate velit esse cillum dolore eu fugiat nulla pariatur.
Excepteur sint obcaecat cupiditat non proident, sunt in culpa qui
officia deserunt mollit anim id est laborum.
Introduction
Regional Setting
Materials and Methods
Results
Discussion
Conclusion
References
Figure Captions
```

The software launches a file browser in which we can select the file, and also launches the *ASCII Filter Options* tool. From these options, we choose the *Western Europe (ASCII/US)* character set, select *Times New Roman* as the default font, *English (USA)* as the language, and CR as a paragraph break, and then save the document as *myfirstpaper_vs1.odt* (see Section 8.4 for more details on character sets, fonts, and paragraph breaks). The default font is *Times New Roman 12 pt*, where Times New Roman is the name of the font and 12 *pt* (or 12 *points*) is the font size. One point is defined as 1/72 inch or 2.54/72 cm, i.e., approximately 0.0139 inches or 0.03528 cm. A larger unit is the *pica* (abbreviated to *p*), which corresponds to 12 pt. Characters in the Times New Roman 12 pt (or 1 p) default font are therefore 12×0.03528 cm $= 0.42337$ cm or about 4.2 mm in height. The Times New Roman font was originally created for the British newspaper *The Times*, in 1931, and its variants are still omnipresent in word processing and book typography.

We may wish to change the *Times New Roman 12 pt* default font to the *Cambria 12 pt* serif font (introduced by Microsoft in the year 2004), or to another serif font. To do this we use *Select All* from the *Edit* menu to select the entire text, and then choose *Cambria* from the drop-down menu of the *Formatting Toolbar*. We could continue editing the text document by, for example, selecting the headers individually to change their font size relative to the main text, or selecting the names of variables within paragraphs to change them from the default text style to Greek characters. Using the *character* and *paragraph* formatting features of the software, however, makes editing the text significantly easier, especially if the style of a large document has to be modified at a late stage in the writing process, or if the document has to be edited by a graphic designer.

Let us now consider the fonts on a computer. A *computer font* is a complete character set of a particular typeface, collected into a *font suitcase* on Apple computers or in the Windows\Fonts folder on PCs. To use a computer font in word processing or desktop publishing software, it must have been installed on our hard drive and made available. Some basic fonts come with a computer's operating system, while other fonts can be delivered and installed with additional software. For example, the now widely used *Cambria* and *Calibri* fonts are included in Microsoft Office and others such as the Microsoft *ClearType* collection are available for free download from the Microsoft webpage; *Minion Pro* and *Myriad Pro* come with the Adobe Creative Suite. Additional fonts of reliable quality containing the glyphs necessary for scientific documents can be purchased from commercial type foundries, such as

```
http://www.adobe.com/type/
http://new.myfonts.com/
http://www.linotype.com/
```

After having been installed, a font will remain as a cross-application unit within a computer's operating system even if the software that it came with is uninstalled, and can be accessed through other software tools. Special *font management* software tools help to install, uninstall, and manage fonts on a computer.

Selecting *Styles and Formatting* from the *Format* menu opens a small window listing the default paragraph and character styles, as well as other styles included in the software. Using *Select All* again and double clicking on the *Text body* paragraph style, changes the font back to *Times New Roman 12 pt*. We notice that the paragraph style has also changed, in that the space between paragraphs has increased by a few millimeters. Checking *Paragraph...* in the *Format* menu indeed shows a 0.21 cm spacing after paragraphs. We change this spacing to 0.00 cm and the line spacing to *Double*, as required by the Guide for Authors for Quaternary Research. We now again change the font of the *lorem ipsum* paragraph to *Cambria 12 pt*. We then choose *Update Style* using the right button of the *Styles and Formatting* window toolbar. This changes the font for the entire text, even though we only changed the font of the *lorem ipsum* paragraph.

We now continue by changing the font of the headers to *Helvetica 14 pt* and *bold*. We select the *Abstract* header and double click on the *Heading* style, which changes the font to the Microsoft alternative to *Helvetica 14 pt*, which is the *Arial 14 pt* font. We then change the spacing above and below the paragraph to 0.00 cm and the line spacing to *Double* using *Paragraph...* from the *Format* menu, and again update the *Heading* style. We can now select the other headers (*Introduction, Regional Setting*, and so forth), and double click on *Heading* to change their style accordingly. Finally, we select the title of the manuscript, change the font to *Helvetica 18 pt* and *bold*, and create a *New Style from Selection* using the right button in the *Styles and Formatting* window toolbar and naming the new style *Title*. To demonstrate how to use character styles, we select the words *labore et dolore* in the text, choose *Variable* from the list of *Character Styles* in the *Styles and Formatting* window toolbar, change the style to *Italics,* and update the style in the same way that we did for the paragraph styles. We have now completed the formatting of the text (Fig. 11.1).

Formatting a large manuscript in this way has the advantage that styles can be subsequently changed in just three simple steps. For example, to change the style of the headings we first select one of the headings, change the font size to 10 pt, and then update the style as we did to update the paragraph styles. We can now see that the font size of all headings has been changed to 10 pt, while the rest of the text remains unchanged. We can modify the *Variable* character style to *underline* instead of *italics* in a similar manner.

This is the title of our manuscript

John Q. Scientist, Jane W. Researcher

University of Nowhere, Department of Science

Abstract

Lorem ipsum dolor sit amet, consectetur adipisici elit, sed eiusmod tempor incidunt ut labore et

dolore magna aliqua. Ut enim ad minim veniam, quis nostrud exercitation ullamco laboris nisi ut

aliquid ex ea commodi consequat. Quis aute iure reprehenderit in voluptate velit esse cillum

dolore eu fugiat nulla pariatur. Excepteur sint obcaecat cupiditat non proident, sunt in culpa qui

officia deserunt mollit anim id est laborum.

Introduction

Regional Setting

Materials and Methods

Results

Discussion

Conclusion

References

Figure Captions

Fig. 11.1 Screenshot of the Lorem Ipsum text after editing the layout with OpenOffice Writer using paragraph and character styles. The default font is Cambria 12 pt, the title is in Helvetica 18 pt, and the headers are in Helvetica 14 pt.

Formatting manuscripts with a commercial text processor such as Microsoft Word or Apple Pages works in an essentially similar manner to that described above, but these software products are slightly easier to use and, at least in the case of Apple Pages, more stable and reliable than OpenOffice Writer. While working on the above example, OpenOffice Writer froze repeatedly when trying to create a new style. Microsoft Word is usually very slow and often crashes, especially on Macs, and the Microsoft User Folder needs to be cleared from time to time to maintain the performance of the software. In any case, when creating large text documents that include numerous objects other than text (e.g., photos, tables, and graphics) it is advisable to use desktop publishing software such as Scribus or Adobe InDesign, rather than OpenOffice Writer, Microsoft Word, or Apple Pages (see Sections 11.3 to 11.5).

CREATING MANUSCRIPTS, FLYERS, AND BOOKS

11

Manuscripts that are to be submitted to journals, however, do not require any further formatting. Journal articles such as the one discussed previously are never submitted as ready-to-be-used print files in the final layout of the journal. The manuscript file instead comprises a simple text file that includes a cover page, the main text, references, and figure captions. Figures and tables are typically submitted separately as vector or raster files. All files are normally submitted electronically via the journal's web interface. As an example, let us assume the following list of files to be submitted online:

```
johnscientist_maintext.doc
johnscientist_mainfigure_1.eps
johnscientist_mainfigure_2.eps
johnscientist_mainfigure_3.eps
johnscientist_supplementarytext.doc
johnscientist_supplementarytable.doc
johnscientist_supplementaryfigure_1.eps
johnscientist_supplementaryfigure_2.eps
johnscientist_supplementaryfigure_3.eps
```

Most journals do not accept OpenOffice Writer or Apple Pages file formats but both of these software products can export text files in the Microsoft Word format with the .doc file extension. Once the files have been uploaded, the software behind the web interface merges the text, figures, and tables into a single PDF document that then needs to be approved by the submitting author prior to final submission. The web interface also asks for input of information on the authors, including their names, affiliations, phone numbers, and email addresses. The submitting and corresponding author is also requested to suggest three to five possible reviewers for the manuscript, excluding colleagues because of possible conflicts of interest.

The merged PDF file of the manuscript then goes to the journal's editorial office where its overall quality and suitability for the specific journal is checked. The editor then contacts two to five reviewers by email and asks them for their availability to review the manuscript. The potential reviewers typically receive the title, the list of authors, and the abstract of the paper to help them decide on their own suitability to evaluate the manuscript. Having agreed to review the manuscript, the reviewers then submit their comments (typically about one page) and a recommendation to accept, to accept with revisions (minor, moderate, or major), or to reject the manuscript. The editor then collects the reviews and decides on the acceptance or rejection of the paper. Once accepted, the manuscript goes to the copy editing office for final layout prior to publication in the journal.

11.3   How to Create Flyers

We will now consider the use of *Desktop Publishing* software to create a *folded flyer*, before later continuing to work on the layout of scientific manuscripts. A folded flyer is usually an attractive, colorful, double-sided product. Flyers are either printed in small numbers (<1,000 copies) on color laser printers and used to introduce research programs at conferences, or they are printed in larger numbers (>1,000 copies) on offset printers and used to promote educational programs at universities, or new commercial products (see Section 11.5 for an introduction to generating PDF files for offset printing).

Whatever its purpose, a flyer should be attractively designed and the information provided easily understood. The text of a flyer needs to be very concise and interesting in order to maintain a lay person's attention. The information provided should be free of complex wording and jargon. The text on a flyer is usually supported by simple graphics and attractive photos, possibly including a simple table summarizing the most important information on, for example, a university curriculum, or a research strategy. The design of a flyer or brochure needs to make it attractive to the intended audience. Flyers often have backgrounds that are either colored or consist of relevant photos, and they are therefore commonly printed on glossy coated paper with a grammage of 135–300 g/m^2. They are commonly made from paper of either US Letter or A4 size which, in the case of a flyer, may then be folded into a *C-fold leaflet*. In print production a *flyer* is the technical term for an unfolded piece of paper. For the product we will create in this section, the technical term in pre-press would be *C-folded flyer, leaflet, C-fold leaflet, gate-fold leaflet* or *3-panel roll-fold brochure*, which are different names for the same product.

Our leaflet will contain the *lorem ipsum* placeholder text from the *myfirstleaflet_text.docx* document. For the leaflet, we will use the following graphics files that were created in Chapters 5 to 7 and subsequently edited in Chapter 8:

```
etopo2_surfaceplotlight_vs5_ps.png
naivasha_image_vs7_ps_cmyk_2500px.jpg
srtm_surfaceplotlight_vs4_ps_cmyk.tif
icecore_lineplot_vs8_ai_fordarkbackground.eps
```

Desktop Publishing with Scribus

In our first example we design a leaflet with the free cross-platform software called *Scribus* (http://www.scribus.net/ and http://docs.scribus.net/). We can either create a new file with *Toolbar > File > New* or use one of the available templates provided by the software, most of which contain a suite of grids and guides. In the *New Document* window, we choose a single-page document layout, set the number of pages to 2, and select an A 4 landscape format with margin guides of 10 mm, bleeds of 3 mm, and millimeters as the default unit for the new blank document. Using *Toolbar > File > Document Setup* we can define numerous presets such as *Document Information, Display, Guides, Hyphenation, PDF Export,* and *Color Management.* With *Toolbar > View* we can toggle the visibility of the document margins, bleeds, guides, rulers and other features. With *Toolbar > Window* we can activate floating panels such as *Layers* and *Action History.* Unfortunately, a clear disadvantage of the software is that we cannot dock or organize the floating panels to arrange them individually within our display. If a second display is available we can arrange our favorite panels there in order to keep them within sight and easily accessible while we work on the document using the first screen.

Using *Toolbar > Page > Manage Rulers* we define vertical guides at 97 mm and 197 mm on the first document page, and at 100 mm and 200 mm on the second document page. These guides represent the fold lines for the leaflet that separate the *cover page* (Page 1), the four *inner pages* (Pages 2 to 5), and the *back page* (Page 6). We then create three text frames on the first page of the document using *Toolbar > Insert > Text Frame* and *Toolbar > Window > Properties* (Fig. 11.2), one with a width of 77 mm and the other two with widths of 80 mm, separated from each other by gaps of 20 mm, which will become pages 5, 6 and 1 of the leaflet after it has been folded. The text frames on the second document page then have widths of 80, 80 and 77 mm, again with 20 mm separation, representing pages 2, 3 and 4 of the folded leaflet.

After linking the text frames 1 to 6 in the correct order using *Toolbar > Item > Link Text Frames,* we import the text contained in the *myfirstleaflet_text.docx* Word file using *Toolbar > File > Import > Get Text,* and starting with text frame 1 on the right of the first document page (Page 1) which will later form the front cover of the leaflet. Unfortunately, the original formatting and line breaks are lost when importing the text and further editing is therefore required. Using *Toolbar > Edit > Edit Text* opens the *Story Editor* window in which we can modify the content and formatting, as well as defining and applying paragraph styles. Using *Toolbar > Insert > Space* we can add frame breaks and line breaks to the document. It is important to note that all changes must be applied to the document using *Toolbar > Edit > Update text Frame* before we save the file as *myfirstleaflet_scribus.sla.*

We then use *Toolbar > Windows > Layers* to create three layers, the upper layer for text, the second layer for artwork, the lower layer for the colored background. To organize objects in layers so that we can toggle

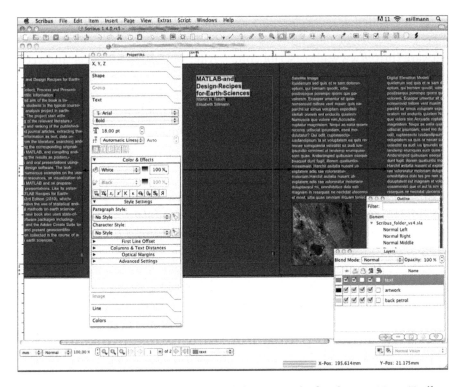

Fig. 11.2 Screenshot of the *Scribus* desktop while creating a leaflet, showing *Menu, Toolbar,* the document itself with a text frame, the *Layers Palette,* and the *Properties Palette.* The latter is the control center of Scribus, useful for many tasks such as defining *Shape, Line* and *Color,* and defining the appearance of the *Text* by applying *Color, Effects, Margins, Paragraph* and *Character Style Settings.*

their visibility, printability, and protection, we use *Toolbar > Item > Send to Layer.* Using *Toolbar > Edit > Colors* we create a new color called *petrol blue* with C = 90, M = 20, Y = 25, K = 70 in the CMYK *Color Model.* We create a rectangle covering both the background and the bleed of the leaflet using *Toolbar > Insert > Insert Shape > Default Shapes,* use the *Properties* window to fill it with the *petrol blue* color, and then move it to the lowermost layer. As we now have a relatively dark background color, we need to change the text color to white and, if we want to we can also change the font size and apply paragraph styles to the text.

Having imported the text we next need to import raster and vector graphics. As an example of a raster image we can import *etopo2_surface-plotlight_vs5_ps.png.* We use *Toolbar > File > Import > Get Image* to import the image into Page 1 (the cover page of the leaflet), and adjust the appearance of the image on the page using the *Properties* window. Importing the vector file called *icecore_lineplot_vs8_ai_fordarkbackground.eps* using *Import > Get Vector File*, however, requires installation of the *GhostScript* software (http://www.ghostscript.com/). After importing all raster and vector graphics, and completing the layout of the leaflet, we generate a PDF file using *Toolbar > File > Export...> Save as PDF...* . In the *Save as PDF* panel, we simply use the default settings and create a PDF file using the PDF/X-3 standard, as recommended in the Scribus documentation.

Desktop Publishing with Adobe InDesign

In this subsection we will work on the same task (designing a leaflet) but this time using the professional and commercial *Adobe InDesign* software (http://www.adobe.com/indesign.html). After launching the software, we set up our workspace using *Workspace > Advanced*. As with Scribus, if a second display is available we can arrange our favorite panels there in order to keep them within sight and easily accessible while we work on the document using the first screen. The Adobe InDesign user interface or desktop is very similar in appearance to that of its sister product Adobe Illustrator, with a menu bar, control panel, toolbox, and multiple panels (such as the layers panel, for example) (Fig. 11.3). We choose *Normal Mode* at the bottom of the *toolbar* to display *non-printing* objects, such as object frames, guides, and a white *pasteboard*. On the pasteboard we can provisionally collect and lay out text fragments, images, and illustrations during the design process. Before we start using the software, we can hover the pointer over the various tools, read the pop-up *tool tips*, and check the drop-down menus in the menu bar. By searching for keywords highlighted in *italic* font in this text, further information, including extensive descriptions of the software and numerous screenshots, can be obtained online from Adobe websites at

    http://tv.adobe.com/

and

    http://help.adobe.com/en_US/indesign/cs/using/index.html

We first create a new document using *File > New > Document* and the *New Document* window. We then choose two single pages in A4 format and landscape orientation. Using *More Options*, we set the margins to 10 mm, the *bleed* to 3 mm, and the layers and guides as described for Scribus in the previous subsection. Alternatively, we can open the template *myfirst-leaflet_template_empty.indt*, which already contains some ready-to-use formats. The first page of the document shows a red rectangle, together with the word *Text*, both underlain by a *smooth shade. File > File Info* opens the *File Info Dialog* box in which we can add metadata (such as a *Description*, the *Author Information*, and *Keywords*) to our document, which will then remain affiliated with the document, and even with the PDF file that will be generated later on, for printing.

We now save the file as *myfirstleaflet_vs1.indd* and check the *panels* of our workspace by clicking on their symbols. In the *Layer Panel* we find three layers called *text, artwork*, and *back color*, with *text* at the top of the list to avoid problems when using transparency, such as *shades* or *filters*. In the *Object Styles* panel we find two default styles marked by square brackets and two new styles called *shadow* and *shadow text*, which we will examine and use later in this subsection. Using *View Grids & Guides > Show Guides*, we find guides for paper folding and lines indicating text margins. We con-

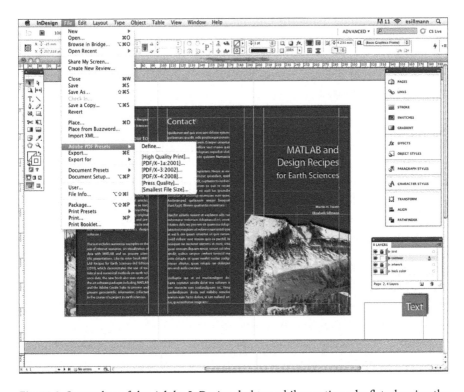

Fig. 11.3 Screenshot of the *Adobe InDesign* desktop while creating a leaflet, showing the *Document* window with rulers, guides, text frames and the document itself, the *Application* bar, the *Control* panel, the *Toolbar*, six *Panel Groups* in the vertical dock and the *Layers* panel. Using *File > Adobe PDF Presets* we can select existing presets or define individual presets for generating a PDF.

tinue to prepare our workspace and open two more panels using *Window >
Object & Layout > Transform* and *Align*. Clicking on the first *Sample Page* in
the *Pages* panel, we find guides representing widths of 97, 100, and 100 mm,
and provisional page numbers 1 to 6 noted on the pasteboard, beyond the
limits of the document so that they will not be printed.

We next explore the two objects that already exist in our document, a
red rectangle and a text box. In desktop publishing text objects are tagged
with *Paragraph Styles* and *Character Styles*, just as they are in word processing. The expression *Text* in the text box is formatted with a *Paragraph
Style* called *Title*. The paragraph style can be defined in the corresponding
panel using the *Paragraph Style Options* dialog box. We use the *Myriad*
font, a regular style, a 30 pt font size, a 36 pt leading, and the color white.
If the chosen font is not available, InDesign highlights the text in red. The
Drop Shadow of the text object is changed by using an object style called
shadow text with an opacity of 90 %, a 1 mm distance, and a –130° angle,
which can be defined using the *Object Style* panel. The red rectangle lies in
the *artwork* layer and its drop shadow is defined by the *shadow* object style.
We choose the *Selection* tool from the toolbox, select the text and rectangle
objects, and either delete them or move them somewhere outside the document on the pasteboard.

Opening the *Swatches* panel shows four default colors *Registration,
None, Paper,* and *Black,* as well as two shades of blue identified by their
CMYK codes. Please note that the color *Registration* is reserved for correlating and aligning the overlapping CMYK colors of the four printing plates.
This particular color has the CMYK values of $C = 100$, $M = 100$, $Y = 100$,
$K = 100$, i.e. all colors are at maximum intensities or 100 %. The registration
marks appear on each of the four printing plates, e.g., as a crosshair target
outside the actual page. While four times full intensity (equal to 400 %) is
acceptable for the registration marks, it is actually too much printing color
on the paper in the offset printing process, which has an upper limited of
300–320 %. On a normal page, black should therefore generally consist only
of $C = 0$, $M = 0$, $Y = 0$, $K = 100$. Another common mistake in offset printing
is the use of *RGB-black*, which consists of full-intensity red, green and blue.
The printer converts the three RGB colors into CMYK colors red, green and
blue, without including key, and black line art, text, and tables therefore appear with blurred outlines or as typefaces of overlapping colors.

We can now create a new color, a dark greenish blue, for the background
of the flyer, which we do in the *Swatches* panel using *Swatches Menu >
New Color Swatch*. We type in the values $C = 90$, $M = 20$, $Y = 25$, $K = 70$ for
the CMYK *Process Color* and click *OK* to add the new color to the panel.

We then open the *Pages* panel, select the *back color* layer, and draw a 300 mm × 216 mm rectangle in the new color. The size of the rectangle also includes a circumferential *bleed*, which is an extra 2 to 3 mm beyond the edge of the 297 mm × 210 mm sheet to avoid any white, unprinted edges in the final, trimmed document.

We next use the *Type* tool from the toolbox to create a text frame in the *text* layer, in between the guides on the cover page. Using *File > Place ...*, we select the placeholder text *myfirstleaflet_text.docx* and place it on the right of Page 1 of the leaflet. When importing the text, we can choose between various options, such as the way in which the styles of the original document are preserved during import. In our example, we decide to retain all the formatting from the word processing software. The original paragraph styles and character styles are automatically converted into the InDesign styles *Title, Header2 18 pt, Header1 24 pt, Normal, text 10 pt folder*, and *Author 10pt bold*. We delete the black RGB swatch in the *Swatches* panel and replace it by *Paper*, which automatically changes the black color into white for all styles.

Continuing with the *Type* tool, we create five new text frames on pages 2 to 6 of the leaflet. We connect these new text boxes with each other by clicking on the *out port* of one box (indicated by a + symbol in the lower right corner of the box), and on the *in port* in the upper left corner of the subsequent box. The text flowing into multiple text boxes can be inspected using *Choose View > Show Text Threads*. Within the flowing text, we can add frame breaks using *Type > Insert Break Character > Frame Break* to divide the text into several segments, each comprising a header and a short text.

Having created the background of the flyer and *placed* the text, we will include four graphics in the *artwork* layer. These graphics (listed below) were created with MATLAB in Chapters 5 to 7 and subsequently edited in Chapter 8 for inclusion in presentations, posters, and manuscripts.

```
etopo2_surfaceplotlight_vs5_ps.png
naivasha_image_vs7_ps_cmyk_2500px.jpg
srtm_surfaceplotlight_vs4_ps_cmyk.tif
icecore_lineplot_vs8_ai_fordarkbackground.eps
```

We use the *Rectangle Frame Tool* to create a rectangular frame on each of the first four pages of the leaflet (Pages 1, 2, 3, and 4), and save the file as *myfirstleaflet_vs2.indd*. Using *File > Place*, we import each of the four figures and scale them to fit within the frames using *Object > Fitting > Fill Frame Proportionally*. We then use the *Selection* and *Hand* tools with the computer mouse, or alternatively the *Direct Selection* tool, to select and then move or scale each graphic object within its frame, without actually modifying the frame itself. To adjust the arrangement and size of the illustrations we can use the *Transform* and *Align* panels. If we wish, we can add drop shadows from the *Object Styles* panel: we can choose between *shadow* object style for illustrations and *shadow text* for text objects such as the title on the cover page.

In our example, we have always used *File > Place* to place objects into the InDesign document instead of including them in the document. The advantage of placing objects instead of including them in the document is that each of these objects can then be further edited using Illustrator or Photoshop, and updated automatically within the Indesign document. Furthermore, when placing instead of including graphics objects the file size of the InDesign document remains relatively small, even if working with high-resolution images. While working on the document in InDesign, the software displays a low screen-resolution version of the image, which facilitates the modification of larger documents.

Working with placed objects requires full control over the various files contributing to the main document. The *Links* panel is used as a tool for managing these files, for navigating to the locations of the various objects included in the document, for opening the *placed* objects (such as PDFs, graphics, and images) with their original editing software, and for updating file paths and versions. The *Links* panel can also be used simply as an information center, providing all details concerning the *placed* objects, the software used for their creation, their date of creation, any use of *transparency*, the location of the objects within the document, and their *paths* (i.e. their locations within the computer's file system).

The illustration *icecore_lineplot_vs8_ai_fordarkbackground.eps* has been specifically designed to stand out well on the petrol blue background, as it has white labels. The graph, however, would look even better on the dark background if the line color was also white and if the area between the two curves was light blue. We can open the illustration in its original software (Illustrator) by first clicking on the figure's name and then on the *Pencil* icon at the bottom right of the *Links* panel. In Illustrator we can then change all the line colors to white, remembering to deactivate the overprinting checkbox in the *Attributes* panel.

We next remove the graph's white background. To do this we open the *Pathfinder* panel using *Window > Pathfinder* while the two curves are still selected, press the *Exclude Overlapping Shape Areas* button, and fill the area between the curves with a new color that has the CMYK code C = 90, M = 20, Y = 25, K = 70 and a 40 % color intensity. After editing the line graph, we save it under the name *icecore_lineplot_vs8_ai_blue_neg.eps* and update the *Links* panel in the main document in InDesign. By using the *content grabber* we can scale and move the line graph within the graphics frame in order to highlight, or zoom in to, a detail of the illustration, provided it is used for decorative purposes only. The content grabber is a particularly useful tool when working on images within the document.

A very nice feature of text frames is the ability to float text around an image using the *Wrap* function of the software. Using *Window > Text Wrap*, or alternatively, by creating an object style for wrapping, we can choose *Wrap Around Object Shape* from the various options available. Another elegant way to wrap text around graphics objects is to manipulate the shape of the text frame using the *Add Anchor Point Tool*. This process is very similar to editing a path in vector graphics software, in that we can add or remove anchor points and then move them around to change the shape of a polygon. After completing the layout of the flyer with InDesign we again save the document as *myfirstleaflet_vs2.indd*.

Preparing the Document for Printing with Adobe InDesign

Having finished our work on the layout of the leaflet, we need to check the document before generating a PDF file. We first examine the list of fonts used in the document, using the *Find Font* dialog box in *Type > Find Font*. In our example document we have used three different fonts: *Minion Regular*, *Myriad Regular*, and *Myriad Semibold*. In order to change a font, or to search and replace for a missing font, we choose *Replace With*, select another typeface, and then click *Find First* to check where the font has been used. We then replace it using *Change All*, with the checkbox *Redefine Style When Changing All* activated. Both paragraph and character styles can be changed with this command, throughout the whole document. Having checked the fonts, we then check the resolution of the file in the *Links* panel. The resolution for a digital printout should be in the range of 150 to 300 dpi. Having checked both the fonts and the resolution of the document, we save it as *myfirstleaflet_vs3.indd*.

To collect and store the main document together with all linked files into a single folder, we use the *Package* feature of InDesign. *File > Package...* creates a new folder with the same name as our main document. Within this folder all files related to our leaflet project are collected into a single location in our file system. The *Package* feature also creates a subdirectory called *links* inside this new folder, gathering together all of the linked files contributing to the leaflet. Having completed the packaging we can finally generate a PDF file using *File > Adobe PDF Presets > High Quality Print* (Fig. 11.4), or any other appropriate choice of a *Print Preset*. When generating the PDF document with color or images running up to the edge of the page, we activate the *bleed* and *registration (trim) marks* options, which are positioned outside the A4 paper format. To ensure that these objects appear on the paper print out, we use an A3 paper format for printing and later cut off the *bleed*. Whereas the *High Quality Print* settings are sufficient for printing the document on a digital laser printer, generating PDF documents for offset printing requires additional settings to be made, such as for *color management, preflight,* and specific *export options* (see Section 11.5 for more details on offset printing).

11.4 Designing a Thesis or a Research Report

The writing and formatting of a multi-page doctoral thesis, a comprehensive research report, or any other long text document, requires more detailed planning than a relatively short, 10 to 15 page journal article. Although there is an increasing tendency in most countries to submit cumulative doctoral theses consisting of several (commonly three) thematically related journal articles, most doctoral candidates submit their theses in edited formats, with graphics objects, tables, and photos included in the page layout. In most cases, text processors such as OpenOffice Writer, Microsoft Word, or Apple Pages can easily manage this task as long as the text documents do not become too large and contain only a few other elements besides the text. This section aims to help us decide on the best tool to use for editing a thesis or research report, i.e. a text document that includes several chapters and sections, with a large amount of text and numerous graphics, tables, images, and appendices.

A popular way to handle large documents using a text processor such as Writer, Word, or Pages is to split them into smaller, more manageable files representing individual chapters. Having finished writing the text, we need to switch to desktop publishing software such as the free *Scribus* or the commercial *Adobe InDesign*. The fundamental difference between word processing software and desktop publishing (DTP) software is a result of their different historical origins. The first word processors were line editors designed for typewriter-style terminals and they therefore had very limited page-making capabilities. In contrast, desktop publishing tools evolved from typesetting printed pages by arranging rectangular objects of text and illustrations within a printing frame.

The distinction between these two ways of creating printed documents is, however, becoming increasingly blurred. Word processors now provide some of the typical desktop publishing features that are included in Scribus and InDesign. For example, Pages by Apple now has both word processing and page layout features, as you can create text boxes that float freely around graphics and images. Of course, these features are very limited compared with those provided by InDesign, but the software easily outperforms Word on both Macs and PCs in its handling, stability, and reliability.

For the user, there are three main differences between text processors and DTP software. The first difference is that, as has been stated previously, both text and images are independent objects in DTP software, organized vertically into *layers* and horizontally within these layers into *frames*. This allows the design of complex page layouts in a workflow that is fundamentally different from that of text processors. In fact, editing page layouts with DTP software has many similarities with vector graphics editors such as Adobe Illustrator. The second difference is that DTP software uses an elaborate option for inserting or *importing* objects such as images, graphics, or PDF files (and many other types of files) by *placing* them in the a document rather than embedding them, as has already been demonstrated in Section 11.3 for flyers and brochures. Placing objects instead of importing or embedding them reduces the file size and can therefore dramatically improve the performance of the software, especially when working with large manuscripts that include a large number of illustrations. As described in Section 11.3, the *path* of the original file stored somewhere in the file system appears in the *Links* panel. This panel is useful for supervising, managing, and updating large numbers of placed graphics objects. The third difference between text processors and DTP software is the consistent application of *color modes* throughout the document in DTP software, and the ability to choose between *CMYK* and *RGB* color modes when using text processors, which is essential when preparing a manuscript for offset printing.

In addition to these three main differences, professional DTP applications offer numerous features that can increase productivity, reliability, and flexibility when creating large documents with complex structures and ambitious layouts. For example, the *layers, master pages, object styles, text styles*, and *indexing* tools, together with the ability to *combine* and *synchronize* multiple documents using the *book function*, are of great assistance to a publisher. Advanced features are available for the creation of publication-quality documents, such as the *preflight* and *production* tools, the handling of *fonts*, the ability to add *bookmarks*, the possibility of using *multiple page sizes*, and support for *PDF/X standards* when generating PDFs. The InDesign software also provides a seamless interface with other members of the Adobe Creative Suite, such as Photoshop, Illustrator, Acrobat, and Bridge.

In this section we do not use the open source Scribus software as it has already been described in detail in Section 11.3. Instead, we use Adobe InDesign to create a typesetting file for a thesis or a research report. Once again, the reader is advised to make use of the comprehensive help and tutorials provided on the software vendor's webpage:

```
http://help.adobe.com/en_US/indesign/cs/using/index.html
```

As in the previous subsection on desktop publishing with Adobe InDesign (in Section 11.3), keywords *highlighted* in *italics* in this section refer to the corresponding Adobe help resources available online.

Authors typically write a thesis or research report on a word processor, rather than using a desktop publishing tool. In this case it is advisable to use paragraph and character styles throughout the document as this will greatly facilitate further processing with InDesign. We have already learned (in Section 11.3) how to create a small leaflet with this software. We again use the placeholder text *lorem ipsum* as an example text for our longer document; this is provided in the Word document called *thesis_chapter_1_ placeholder_text.docx* and uses paragraph and character styles. The Word file does not contain any graphics, as these are provided in a separate folder for typesetting. The document instead lists all figure and text captions at the end of the text. The list refers to graphs and images edited in Chapter 8, which we will also use in our thesis or research report. The following tutorial creates a document for digital printing in a single-sided layout using InDesign. Later on (in Section 11.5) we will use the same material but change the manuscript into a double-sided book for professional printing.

Creating a New Document with Layers, Master Pages, and Pagination

After launching InDesign, *File > New > Document* opens the *New Document* window where we create eight A4 pages with a portrait orientation. To define the *type area*, we choose *More Options* and set the top and left margins to 25 mm, the bottom margin to 15 mm, and the right margin to 20 mm. Since most theses are printed using digital laser printers we do not need to include *bleed* areas in the document as we did for the leaflet in Section 11.3. We then create four layers called *text*, *artwork*, *figure caption*, and *running head*, from top to bottom. Finally, we add metadata such as the author information, the title of the document, and so forth, to the file using *File > File Info*. If required, we can set up a suitable workspace with *Workspace >*

Advanced, as described in the previous section.

Having completed the setup of our DTP document, we then proceed to work on the individual pages. The *Pages* panel displays eight pages and a single A-Master page. Selecting the *Master* page A opens the corresponding master page, on which we can work as on a regular page, but all modifications and additions will then feature on every page of our document to which this particular master page is applied. Using the text tool we create a flat text frame for the *running head* element, immediately above the print area in the corresponding layer, and type a placeholder text such as *The Title of my First Thesis*. For this short piece of text we create and apply a new *Paragraph* style. We call it *running head*, open its *Paragraph* style in the *Paragraph* style panel, define the *Basic Character Formats* as *Font Family* Myriad, and set the *Size* to 10 pt, the *Leading* to *Automatic*, and the *Case* to *All Caps*. Changing the *Case* to *All Caps* turns all characters into capital letters. To create a dividing line we use *Paragraph Rules*, click the *Rule Below* checkbox, set the *Weight* to 0.75 pt, and choose a black solid line of the same *Width* as the column. We set the *Language* to English USA, the *Indents and Spacing* to 15 mm *Left Indent*, and the *Alignment* to centre.

For automatic pagination of the document we create a small text frame in the upper right corner of the master page, choose *Type > Insert Special Character > Markers > Current Page Number*, and insert a text variable for the page number. We duplicate the *running head* style, change the font style to *Semibold,* deactivate the *Rules* checkbox, set all indents to 0 mm, set the *Alignment* to left, name the new style *pagina*, and align the running head and pagina with each other. Scrolling through the normal pages in the *Pages* panel or document window shows that all pages now have a running head and a continuous pagination.

Importing Text, Using Text Frames, and Using Flowing Text

Having created a new document with layers, master pages, and pagination, we will now place text into the text frames. Using the *Type* tool we create a text frame on the first page, flush with the page margins. If necessary, we can use the *Selection* tool to adjust the size of the frame and move it, using the frame handles. We then activate the text frame and use *Menu > Place …* to select *thesis_chapter_1_placeholder_text.docx*, which contains the text to be placed in the document. In the *Show Import Options* dialog window, we choose to import the text together with the text format. Using *Open*, we import the text into our typesetting file. We find all paragraph styles marked with a small floppy disk icon, indicating that all new styles need to be checked; we will return to this topic in the next subsection.

Having imported the text we notice that only the first few paragraphs are actually visible because the text is larger than the text frames on the first page. This *overset text* requires the creation of additional text frames, linked to the first text frame. We click on the red plus-symbol inside the *Out port* in the lower right corner of the frame and then use the *Selection* tool to show the *loaded text icon*. We can drag out a second text frame on the second page and again fit it to the page margins. Alternatively, we can use the *Autoflow* feature of the software to create multiple text frames on many pages by holding down the *Shift* key while creating the next frame on page three. Using *Choose View > Extras > Show Text Threads* we can visualize the links between the text frames (Fig. 11.4). The *Pages* panel now lists eight document pages, of which seven contain text. We save a version of our file as *myfirstthesis_chapter1_vs1.indd*.

When importing text we try to keep all the original fonts and styles of the Word document. Using *Type > Find Font* we examine the list of fonts for the document: a yellow triangle indicates any fonts that are not available on our computer and therefore need to be replaced. Activating the *Redefine Style When Changing All* checkbox and choosing *Change All* updates all paragraph and character styles, as well as eliminating any font conflicts within the document – an important feature to prevent serious problems during printing. Redefining the styles actually ensures that all fonts will be embedded into the later PDF file and avoids unpleasant surprises when having the PDF printed in a commercial print shop. This contrasts with the clandestine replacement of fonts by word processing software, instead alerting the user if there are missing fonts.

Alternatively, we can use the *Find Font* feature to replace fonts throughout the document without having to modify each individual paragraph and character style manually. This is very useful if, for example, we wish to switch from one variant of the *Times* font family to another (such as from *Times* to *Times New Roman*), or if we wish to make a more general change from a *serif* font to a *sans-serif* font. We can also use this feature to alter the appearance of text fragments from different parts of a cumulative thesis, or of the contributions made by different members of a research project to the final report. In our example we change all fonts into a sans-serif typeface such as *Myriad*, with the *Italic, Regular,* and *Bold* style types. All style definitions and all text in our document are adjusted simultaneously.

Direct Formatting, Paragraph Styles, and Character Styles

Each InDesign document has a *Basic* paragraph and character style that is applied to unformatted typed or pasted text. We can define and modify paragraph and character styles in the same way as with word processors, choosing the font family, style type, size, color, alignment and other attributes, in the *Paragraph* and *Character* panels or in the *Control* panel. An easier and more thorough way to define automated formats is to use InDesign's centralized attributes called *Character Styles* and *Paragraph Styles*. Clicking into the text and then on the *Create New Style* icon at the bottom of the panel opens the *New Paragraph Style* window in which the formatting is already compiled as the *Style Settings* and where we can give a name to the style. To the left we find a list of more options for fine adjustment of the document's format. To apply the text format to a new paragraph

Fig. 11.4 Screenshot of the *Adobe InDesign* desktop while creating a thesis, showing a figure and its caption provisionally prepared on the pasteboard. The text frame on the document page is already reduced to a smaller height where space for the figure is needed, letting the threaded text flow between the interconnected text frames.

we simply click into a text paragraph and then choose the desired style from the paragraph style panel.

Let us now take a closer look at the existing styles in our document, which we imported together with the text. By selecting the red text on the first page, the associated *normal* paragraph style and the *comment in red* character style are highlighted with light blue in the corresponding panels. Clicking the highlighted character style opens the *Character Styles Options* dialog box. The *Basic Character Format* is Myriad Pro Regular and the *Character Color* has no stroke but a fill color. As is typical with word processing software, however, the fill color is in the RGB mode, which is indicated by the three-colored rectangle to the right of the color called *RTF r190 g0b0*. We change this color into a CMYK color with C = 15, M = 100, Y = 100, K = 0. We notice that only a small number of attributes differ from the *normal* paragraph style and therefore delete the Word style and replace it with the InDesign *normal* paragraph style. The style called *figure caption bold* is still in the panel and still shows the floppy disk symbol, but it disappears after clicking the style and checking the settings in the dialog window. The original font of, for example, labels such as *Figure 1.1*, *Table 1.1* in each of the figure or table captions at the end of the document, is changed to *Myriad Pro Bold*, agreeing with the corresponding *Paragraph* style.

We will next edit the style for bullet point lists such as the one after the red *Here comes a list of four bullet points* comment in the placeholder text. Clicking into the list, the paragraph style *normal bullets* is highlighted in the *Paragraph Styles* panel, indicating the style that is already applied to the list, even though the list does not yet have any bullet points. Opening the options, we switch the *List Type* from *None* to *Bullets* in the *Bullets and Numbering* selection and choose a dot as the marker from the *Bullet Character* display. Should the *Bullet Character* display still be empty, we choose *Add* to open a new dialog box and select the *Myriad Pro* font and a symbol suitable for use in scientific texts from the glyph list. Using *Text After*, we insert a *Tab* and set the text alignment to left, the indent to 15 mm, and the tabulator position to 21 mm. In the *Indent and Spacing* options we set the *First line indent* to 10 mm. To check the results we click on the *Preview* checkbox in the lower left corner of the dialog box so that we can toggle between old and new definitions.

In addition to these basic modifications to the document, we can introduce automated numbering of the paragraph styles as *heading 1*, *heading 2*, and *heading 3*. We can do this by choosing *Numbers* as the *List Type* in the *Bullets and Numbering* options, starting a new list, and setting the *Mode* to *Start at 1* for the highest level, i.e. *heading 1*. We can check whether our automated numbering works correctly since the previous numbering is still present as plain text.

We save our file as *myfirstthesis_chapter1.indd*, and also save an additional copy of it as *myfirstthesis_chapter1_vs1.indd*, overwriting the preliminary first version of the file.

Creating Book Files, Page Numbering, and a Table of Contents

This section is on examples of what are known as the *Long document features* of desktop publishing tools. The first example of these features that we will explore is the InDesign book function, which helps to make large publications manageable. We first create a new empty document called *frontmatter* containing all settings used in the other chapters of the thesis or research report. We open *myfirstthesis_chapter2.indd* by clicking the file on the book panel, save it as *myfirstthesis_frontmatter.indd*, and remove all text, figures, and captions. We import all text contained in the Word file *thesis_frontmatter_placeholder_cover.docx* and place it in Page I and and then import all text in the Word file *..._abstract.docx* and place it in a separate text frame on Page IV. We then duplicate and modify the existing paragraph styles to format the cover page and save the file as *myfirstthesis_frontmatter.indd*.

We next use *File > New > Book* to create a new file and save it as *myfirstthesis_book.indb*. The book panel appears and we can add new chapters by clicking on the *plus* icon in the lower right corner of the panel. We then add three existing documents created before to the new frontmatter and Chapter 1 documents:

```
myfirstthesis_chapter2.indd
myfirstthesis_chapter3.indd
myfirstthesis_backmatter.indd
```

In the book panel's *Document Numbering Options* context menu we choose Roman figures *I, II, III, IV* … starting with the first page of the frontmatter, to keep the pagination independent of the main content of the thesis. For the first chapter of the book, we again start with page number 1 but now use Arabic figures *1, 2, 3, 4*…. For the subsequent chapters and the backmatter we again select Arabic figures, but then choose *Automatic Page Numbering*, continuing from the previous book document (Fig. 11.5).

We then open the frontmatter document and adjust the contents settings using *Layout > Table of Contents* to create a preliminary version of the *Table of Contents* (TOC). In the right column, we click on *<< Add* to add the *header 1*, *header 2*, and *header 3* styles to the *Include Paragraph Styles* list on the left. We then click the button *More Options*, select the *Include Book Documents* checkbox, and use the *content 1*, *content 2*, and content 3 paragraph styles. We adjust the paragraph style options for each entry and, in addition, use the *dots content* character style to format the table of contents automatically. After generating the TOC, a text cursor pops up that enables us to create an independent text frame flowing onto pages II and III. Checking all entries we find a typing error in *Palceholder with Typo for Header 2*. We do not edit this error in the TOC itself but instead open the first chapter, correct the typo there and then update the TOC. In the next subsection we place images and a table into Chapter 1 and then again update the TOC.

Fig. 11.5 Screenshot of the *Adobe InDesign* desktop while creating a thesis, showing the first page of Chapter 3, with text formatted using three levels for the headers and body text with the formats *normal*, *indent* and *normal bullets*. The running head on the first page and pagination of a new chapter still has to be removed. The *Book* panel shows all parts of the book listed, with *Automatic Page Numbering* activated and Roman page numbers for the frontmatter and Arabic page numbers for the other pages of the book. The red and green bullets in the panel indicate the result of the *preflight* for each chapter.

Placing Images and Tables, and Adding Captions

Figures and tables are usually placed at the top or at the bottom of a page, after having been discussed in the text. We can use the *Find/Change* function to search for *fig* and *table* to find any references to figures and tables within the text. We find that *Figure 1.1* is first mentioned on page 2 of the document. We then click on the text frame in which the figure is discussed and reduce the height of the frame, thus moving some of the text to the next page. We then find that *Figure 1.2* is mentioned on page 3 of the document and *Table 1* on page 6, so we also modify the sizes of the frames on those pages in a similar manner. We provisionally place the following graphics on the pasteboard outside the actual document (as described in Section 11.3), in a layer called *artwork*, to be later moved into the blank spaces that we have created in the document:

```
icecore_lineplot_vs5_ai_cmyk.eps
etopo2_surfaceplotcontours_vs3_ps.jpg
```

We again either place or import the graphics as we did in Section 11.3 (and for the same reasons) rather than embedding them into the document using *copy and paste*. This approach allows all files used in the document to be subsequently collected into a single folder using *Choose File > Package*, no matter where on our computer they were stored originally. We again use the *Links* panel to manage the placed figures.

When working on a manuscript we need to have access to the names and *Status* of each of the graphics, and to additional information such as their *Page* number, *ICC Profile*, *Effective PPI* and *Transparency*. The status of graphics is shown as *Up to Date*, *Modified*, *Missing*, or *Embedded*. If the status needs to be modified, the *Links* panel is quite helpful in finding, relinking, and editing graphics. When updating the status we can also adjust the preview resolution for the graphics as they are displayed on our computer screen using *View > Display Performance > High Quality Display* or … *> Fast Display*, depending on our preference. As stated previously, the figure and table captions are normally included at the end of the document. We can copy them from there and paste them into individual new text boxes in the *figure caption* layer. We can then give them the *figure caption* object style and, if necessary, resize them to fit into the frame.

Using *File > Place*, we import the *geochem_data.txt* file that was edited as tab-delimited text in Section 8.5. Alternatively, we can just copy and paste the content from a word processor or spreadsheet software file. Using *Table > Convert Text to Table*, we convert the tab-delimited table into a formatted table by choosing *Tab* as the column separator and *Paragraph* as the row separator. We create the *table head* paragraph style with *12 pt bold* characters and the alignment set to *Centre*. We use the same *12 pt regular* font as in the main document, set the alignment to *Right*, and set a *Right Indent* of *17 mm*. Using *Table > Table Options* and *Cell Options* we give the table border and strokes a *1 pt* thickness and set the border and strokes color to *Paper*. We set the fill color of the body cells to *Blue 15 % Tint*, and the fill color of the header row to *Blue 30 % Tint*. As was the case in Chapter 4, the absurd value of *–999* indicates missing data in the table. We replace the short hyphen for the minus sign by the special longer hyphen or so-called *En-Dash* character using *Type > Insert Special Character > Hyphens and Dashes > En-Dash*. We then save the first chapter of the thesis or research report as *myfirstthesis_chapter1.indd*. We must remember to update our TOC using *Layout > Update Table of Contents* in the frontmatter document, which is listed in the *Book* panel. We then save this file as *myfirstthesis_frontmatter.indd*.

Preflight, Packaging, and Generating a PDF File

The *Book* panel is an efficient control center when completing large documents, being used to check all technical aspects of the files and to generate the PDF file for printing. The panel can be used to *synchronize* or *manage book* documents, *update* their numbering, and to *preflight*, *package*, and *export* the elements of the book.

Before passing on the files for printing it is essential to perform a quality check on all of our files. This so-called *preflight* simulates the printing process by searching the entire book for inconsistencies such as missing fonts, status problems with linked illustrations, low-resolution images, overset text, and so forth. Whenever a red bullet appears next to a file in the *Book* panel, it indicates that an error has been detected, whereas a green bullet indicates no errors in the book file.

The *preflight* feature generates a report on the results of the search. We can launch the *Preflight* panel using *Window > Output > Preflight*. After fixing all existing problems (as described in the subsection entitled *Preparing the document for printing with Adobe InDesign* in Section 11.3) we use *File > Package*, check once again the summary of all items included in the book, and then save everything in new folder which we name *11_4_myfirstthesis_book_folder*. As mentioned in Section 11.3, packaging creates a directory inside this folder called *links*, in which all linked graphics files are stored. This is very useful if numerous files that were previously spread all over the file system on your computer are now collected together into the *links* folder, as this makes it easier to hand over the book files to another person, or to generate an updated PDF file after modifying the typesetting file.

The final step in this subsection is to create an Adobe PDF file, either using *File > Export* or by creating it directly from the *Book* panel using *Adobe PDF Presets > High Quality Print* (or any other appropriate *Print Preset* for digital or desktop printing). In addition to the print file, we can create a second, smaller version of the file suitable for publishing online or sending by email. We can accomplish this by using a different, very convenient *Adobe PDF Preset* such as *Rich Content PDF*, which includes interactive elements, or the *Smallest File Size* feature, which reduces the resolution of all images. In the professional Adobe *Acrobat* PDF software we can examine our PDF file, fill out the *Description* form, set the *Initial View* of the document in the *Page layout* to *Single Page*, and set the *Magnification* to *Fit Page*. If the thesis or research report is to be printed in large numbers, offset printing is preferable to digital laser printing. In Section 11.5 we will demonstrate how to lay out a double-sided product, again using InDesign's *Book File* feature. The more technical aspects of printing large numbers of copies, such as the use of *color management* and the relevance of specific *export options* for generating *PDF/X-1a standard*, high resolution, press-ready PDFs, are also described in Section 11.5.

11.5 Assembling and Laying Out Books

Printed books are generally paperback or hardcover multi-page text documents, typically with more than two or three hundred pages, either in black and white or including colored graphics and images, sometimes with CDs or DVDs attached. The readership of a science book may range from undergraduate students looking for an introduction to a specific topic or wanting to learn about a new method using a tutorial-style textbook (such as the one you are currently reading), to experts seeking information on the results of a major research program.

During the past 5 to 10 years a new product, the *electronic book* or *e-book*, has emerged from a former niche product to a widespread medium, paralleling the rapid development of computers, personal digital assistants (PDAs), smartphones, electronic book readers (such as Amazon's *Kindle*), and other electronic devices on which e-books can be purchased, stored, and displayed. The boundaries between classical publication formats, such as those used for journals or books, and online media used to present scientific content, are consequently becoming increasingly blurred. In such publications, graphics and audio-visual elements that one usually expects to be included on Internet websites (such as video clips, interactive elements, and hyperlinks) can be embedded into a fixed layout format *PDF* or a floating layout format *ePUB* (short for *electronic publication*). Both text processors and DTP software offer built-in converters to create such ePUB files, but publishers now prefer cross platform *XML*-files. Printed editions may try to compensate for this added value by providing supplementary data on a CD or DVD, or by making it available online through a companion website.

Assembling a publication, whether for publication online or in a traditional paper format, typically starts with the meticulous work of collecting and creating content, drafting an outline, writing the main text, compiling tables, listing references, and designing figures. In order to avoid unpleasant surprises when offset printing the final product or publishing an ePUB document online, most authors submit raw text and graphics material rather than providing the publisher with a print-ready file. The publisher's webpage usually provides the authors with guidelines for organizing the material into an appropriate format, for example in *Book Manuscript Guidelines*, *Author Instructions*, or even a *Microsoft Word document template*. Springer, for example, provides concise information at

```
http://www.springer.com/authors/book+authors?SGWID=0-154102-0-0-0
```

as well as providing a ready-to-use *Word* template and *LaTeX2e* macro packages at

```
http://www.springer.com/authors/book+authors?SGW
ID=0-154102-12-417900-0
```

We can download the *T1-book.dot* file from the Springer website and use it as a template for laying out new text. After activating the *TextTools* and *TitlePage* toolbars using *View > Toolbars*, we can either type new text into the template, or copy and paste a text fragment from another text editor into the file. The template contains various predefined *paragraph styles* and *character styles* such as *H1, H2, H3* for headers, typographic alignment styles such as *Justified* and *Indent*, and the *Index Entry* tool for including keywords in the book's index. As already stated in Section 8.4, strict use of *paragraph styles* and *character styles* from the very beginning when writing a book simplifies the layout process significantly, including the compilation of a table of contents and an index or conversion to *XML*.

In all earth science publications, high quality figures are essential. Perfectly reproduced illustrations attract attention to the publication and its author, as well as adding refinement to a scientific work. A publisher's *Artwork and Illustrations Guidelines* provide concise technical information for preparing and submitting figures electronically, such as:

```
http://www.springer.com/authors/book+authors?SGW
ID=0-154102-12-417900-0
```

Color and Color Management

As demonstrated in Chapter 8, the choice of an appropriate *Color Management* such as, for example, the definition of an appropriate color profile for the publication, is essential if a high quality product is to be achieved. General information on color profiles and a detailed description of available color profiles can be obtained at

```
http://www.color.org/
```

for the US color standards and

```
http://www.eci.org/
```

for European standards. Adobe also provides its own information resources, such as

```
http://help.adobe.com/en_US/indesign/cs/using/index.html
```

where we can choose *Color* to access all relevant information on working with colors using Adobe software products. The fundamental problem with color is to distinguish between the use of the *RGB color mode* for computers, projectors and digital cameras, and the *CMYK color mode* for four-color (4C) printing. For example, most digital photographs include a standard red, green, blue (*sRGB*) color profile. The color modes and profiles of photographs should not be modified during processing unless the final output format (printed book, e-book, or online publication) and its specific color management settings have already been defined. For example, an incorrect choice of a CMYK color mode may result in large files with pale colors when published online, while RGB photographs typically appear dark and grayish when printed in a book.

Compilation of the Manuscript Files and Submission to the Publisher

Authors usually submit their book manuscripts to a publisher either on a CD or DVD, or online via the publisher's FTP server. New *book proposals* are usually made submitting a *sample chapter* with a single PDF file that includes all text, figures, and tables. Once the proposal has been accepted, the manuscript is delivered in an editable open-format file to the publisher, for example as an RTF or a Word document. The additional PDF version of the manuscript needs to be carefully checked for the correct appearance of all special characters and fonts, in particular where they occur in figures. In most programs *File > Print* or *File > Export* opens the print dialog box in which we can choose *Adobe PDF* as the export format and then click *Save* in order to transform the manuscript into a PDF, which the author can then approve prior to submission. When submitting multiple files rather than a single file into which all files have been merged, the file names should be carefully chosen in order to allow the editor to associate each file with the relevant book chapter, figure, attachment, and so forth. Below is an example of a hierarchical structure used for the compilation of files to be submitted to a publisher:

```
miller_booktitle_year
   miller_chapter_1
      readme_ miller1.doc
      miller1.doc
      miller1.pdf
      miller1_fig_1_1.tiff
      miller1_fig_1_2.eps
      miller1_fig_1_3.eps
   miller_chapter_2
      readme_ miller2.doc
      miller2.doc
      miller2.pdf
      miller2_fig_captions.doc
      miller2_fig_2_1.eps
      miller2_fig_2_2.tiff
      miller2_fig_2_3.eps
      miller2_fig_2_4.tiff
   miller_chapter_3
      miller2.doc
      miller3.pdf
```

In this example, *miller_booktitle_year* is the name of the folder containing the entire book. The folders *miller_chapter_1* to ..._3 contain all files comprising the chapters 1 to 3, the optional *readme* files provide supplementary information on the contents of the folders, the *.doc* files are the text files for the chapters (including the figure captions). The *.pdf* files are preliminary PDF versions of the chapters (including text, tables and figures). All other files (such as the *.eps* files and *.tiff* files) are the figures, with the relevant chapter number included in the file name.

Working on a Book Project with InDesign

This subsection considers some of the advanced features of InDesign that are particularly useful when creating books for offset printing, such as the *color management* and *preflight* features, and some specific *export options*. Please note that we use *Chapter* and *Section* (with upper case *C* and *S*) to refer to chapters in the book that you are reading, and *chapter* (with lower case *c*) when referring to chapters of the theoretical book being discussed in this demonstration. We will use the text and figures in *myfirstthesis* from Section 11.4 to create a double page spread book. We first create six new folders for the frontmatter, backmatter, and the four chapters of the thesis. Each of the four folders of the thesis chapters contains a *links* folder, which we copy and paste into the equivalent folders of the new book.

```
sjq_myfirstbook_a_frontmatter folder
sjq_myfirstbook_chapter_1 folder
sjq_myfirstbook_chapter_2 folder
sjq_myfirstbook_chapter_3 folder
sjq_myfirstbook_x_backmatter folder
```

Within each *links* folder we add the author's initials *sjq* (for Scientist, John Q.) and a descriptive or *aptronym* part to the figures' file names. Below is a list that compares new file names used in this section with the old file names that were defined in Chapter 8 and then used in Section 11.4 (in brackets), starting with those in the *links* folder for chapter 1 of the new book,

```
sjq_myfirstbook_fig_1_1_icecore.eps
(icecore_lineplot_vs5_ai_cmyk.eps)

sjq_myfirstbook_fig_1_2_surfaceplotcontours.jpg
(etopo2_surfaceplotcontours_vs4_ps_1780px.jpg)
```

followed by those in the links folder for chapter 2,

```
sjq_myfirstbook_fig_2_1_filledcontourplot.eps
(coastline_linegraph_vs3_matlab_4426vertices.eps)

sjq_myfirstbook_fig_2_2_srtmsurfaceplotlight.jpg
(srtm_surfaceplotlight_vs4_ps_cmyk.jpg)

sjq_myfirstbook_fig_2_3_coastlinelinegraph.eps
(coastline_linegraph_vs3_matlab_4426vertices.eps)

sjq_myfirstbook_fig_2_4_srtmfaultssatmerged.eps
(srtm_faults_sat_merged_vs4_ai.eps)
```

and for chapter 3,

```
sjq_myfirstbook_fig_3_1_etopo2surfaceplotlight
(etopo2_pseudocolorplot_vs3_ai.eps)

sjq_myfirstbook_fig_3_2_rosediagram.eps
(directional_rosediagram_vs2_ai_fill1.eps)

sjq_myfirstbook_fig_3_3_etopo2pseudocolorplot.eps
(etopo2_pseudocolorplot_vs3_ai.eps)
```

We then open a new empty InDesign document using *File > New Document* and set up four *Facing Pages* with B5 *Page Size*, which has a *Width* of 176 mm and a *Height* of 250 mm. The margins of the page are set to 25 mm for the *Top*, 15 mm for the *Bottom*, 25 mm for the *Inside*, and 20 mm for the *Outside*, with a *Bleed* of 3 mm. We can now import the text contained in *thesis_chapter_1_placeholder_text.docx*, as described in Section 11.4. Activating *Import Styles Automatically* in the *Import Options* ensures that the text keeps its formatting (such as its paragraph and character styles). For the other components of the book (such as the frontmatter, chapters 2 and 3, and the backmatter) we can use draft InDesign documents with file names that end with *_draft_vs1.indd*. We can import all styles from one of these draft documents into our new document, and then create a new *Book* file called *sjq_myfirstbook_draft.indb* in which to compile and manage each chapter individually. The book uses relatively small font sizes compared to the thesis, such as 20 pt, 16 pt, and 12 pt for headers 1, 2, and 3, 10 pt for the main text, and 9 pt for the figure captions.

For chapter 1 we can also use an InDesign template, such as *sjq_myfirst-book_empty.indt*, which creates a new empty document called *Untitled-1*. The original template with the file extension *.indt* remains unchanged, and we now can either place and import the placeholder text, figures and captions from the *links* folder into the new document. We need to search any missing fonts and replace them with fonts available on our computer by choosing *Redefine Style When Changing all* and *Change All*. Opening the *Pages* panel shows a double paged document, starting with a single right-hand page (page 1), three double pages (left- and right-handed pages 2 + 3, 4 + 5, and 6 + 7), and ending with a left-hand page (page 8).

We open the *Master* page by clicking the small preview icon in the *Pages* panel. In the *Master* page we can add and modify items such as running heads, pagination, text frames, and the file modification date in the center of the double page, all of which are located in separate layers. The text modification date in the *DRAFT* layer, for instance, can be modified by using *Choose Type > Text Variables > Define* and then *Type > Text Variables > Insert Variables > Modification Date*. The file modification date is then automatically updated when opening and saving the document. We can use another *Text Variable* to create *dynamic column titles*, which generates column titles and automatically updates them when the document is modified. For this we define a *Text Variable* referring to *header 1* in its last occurrence on a page, and change the settings of the variable to *running head* on the right page in the *Master* page. We save our file as *sjq_myfirstbook_chapter_1_draft_vs1.indd*, which contains text only at this stage of our work, but we can gradually place figures and captions and save the intermediate steps as versions labeled from ..._draft_vs2 to ..._draft_vs5.

Another advanced feature when using InDesign to create books is the *Baseline Grids* feature. Baseline grids help us when adjusting the text to line up the columns on a single page, on two facing pages, or between the two sides of the (slightly transparent) pages of the book. We can explore the default baseline grid in our document using *View > Grids & Guides > Show Baseline Grid*. We can click within an arbitrary text paragraph and then on the *Align To Baseline Grid* button in the *Control* panel. To adjust the text to the grid for the entire document, we open the *Paragraph Style* panel, click on the *normal* style, select *Indents and Spacing* in its *Paragraph Style Options*, and activate the *Align to Grid All Lines* option.

The two other styles used for the main text that require adjustment are the *normal indent* and *normal bullet* styles. Using *Edit > Preferences > Grids* (for Windows) or *InDesign > Preferences > Grids* (for Mac OS) we keep the default *Light Blue* color for the grid, let it start at 78.5 pt *Relative to Top of Page*, and select *Increment Every* 12 pt (because the body text of our document has a 12 point leading, as defined in the *Basic Character Formats* of the *Paragraph Style Options*). For a perfect adjustment of the headers 1, 2, and 3, we specify their *Leading, Space Before,* and *Space After* values such that the sum of all values equals 12 pt or a multiple of 12 pt. As an example, we use Leading = 12 pt, Space Before = 24 pt, and Space After = 12 pt for header 1, values of 18 pt, 18 pt, and 12 pt for header 2, and values of 16 pt, 12 pt, and 8 pt for header 3. Having completed our work on the baseline grid we save the document as *sjq_myfirstbook_chapter_1_grid.indd*.

The last advanced editing feature of InDesign that we will consider (which is also available with most word processing tools) is the ability to use non-breaking or hard spaces to prevent any line break, for instance, between numbers and units, or between *Fig.* and the number of the figure. We can use *Type > Insert White Space > Nonbreaking Space (Fixed Width)* to insert a non-breaking space. With *Fig. 1* and other figure numbers, for example, which may occur many times within a text, we can also use *Edit > Find/ Change*, type *Fig.* followed by a blank space into the *Find What* box, and then replace this string by *Fig.* followed by a *White Space > Nonbreaking Space* with a constant width.

Having completed the drafts of the chapters, we can create the new book file *sjq_myfirstbook_draft.indb* in which to compile and manage all files. We then place our figures and captions, format the table, and create a table of contents (TOC) for the frontmatter, as described in Section 11.4. We save copies of the book chapters and define names such as *..._draft_vs2.indd*. We then generate a PDF file called *sjq_myfirstbook_draft.pdf* using *Export Book to PDF* from the *Book* panel menu, choosing the *Smallest File Size* option in order to create a small file size suitable for emailing to the proofreader.

The proofreader (or any other colleague working on the document) edits the PDF file using Adobe Acrobat (http://adobe.com/acrobat). After launching the software, we open *sjq_myfirstbook_draft.pdf*. We can use *View > Page Display > Two-Up* to change the page display to *Show Cover Page During Two-Up*, and browse the document pages. Using *File > Properties > Description* we fill in the *Metadata*, which contains a set of relevant information about the document, such as *title*, *author*, and some *keywords*. Using *Choose View > Toolbars > Comment & Markup* we can launch the *Commenting and Markup Tools* to add *Sticky Notes*, to *Cross Out* or *Highlight* text, or to draw geometric shapes such as lines or arrows within the document. In the event that our proofreader (or colleague) does not have access to the professional version of Acrobat, the free *Adobe Reader* software can also be used for marking PDF files if we have activated *Comments > Enable For Commenting In Adobe Reader* in Acrobat before saving the file. After receiving a commented PDF file we can click on *Show Comments List* in the *Comment and Markup* toolbar to display all comments and a toolbar for *Managing Comments*, which includes functions such as *sorting, filtering*, and *replying to comments*. The author or designer can then modify the original InDesign file, and create a new PDF file.

Generating the Final PDF File for Printing

After the proofreading, we save a fresh set of folders and files for the document using *File > Package* (this time without affixing the word *..._draft*) which we call *sjq_myfirstbook.indb*, *sjq_myfirstbook_chapter_1.indd*, and so on. We clear up all book documents, remove all elements from the *pasteboard* and from within the document (such as in the *DRAFT* layer) as they are no longer needed, and check that all the fonts used in the book are available. We can then run a *preflight* for the whole book in which a printing process is simulated by the software and may yield a number of error messages indicating different kinds of problems with the file (such as missing fonts, for example), but ideally responds with green bullets in the panel indicating a successful preflight with no problems encountered.

We then check all colors in the file. Opening the *Swatches* panel of chapter 1, we find four default color symbols listed: *None* (for no color), *Paper*, *Black*, and *Registration*, the latter only being used for technical purposes such as for crop marks in the document. The blue color C = 80, M = 45, Y = 20, K = 15 is used for headers and the table background, and K = 30 (or 30 %) black is used for the file modification date in the center of the double page. Importantly, we notice that there is a single red color that resulted from the text import from the Word file. This is an RGB color, as indicated by the three-colored square to the right of the panel, and therefore not suitable for offset printing. We therefore select this color, which is called *WORD_R190_G0_80*, and all other superfluous colors in the panel, click the *Delete* icon at the bottom of the *Swatches* panel, and use *Remove and Replace* to remove the red color and replace it with the default *Black*.

Using *Window > Output > Separations Preview* produces a list of five rows in an independently floating panel. The first row shows a combination of the four CMYK *printing plates*, while the other rows each show only a single plate for *Cyan*, *Magenta*, *Yellow*, or *Black (Key)*. We wish to use only *process colors* in our document, these being combinations of *Cyan*, *Magenta*, *Yellow*, and *Black (Key)*. All other colors are converted into the four process colors using the *Swatch* panel. When toggling the *visibility* of the black plate, black text, or outlines, we notice that the main text, the strokes, illustration labels, and all other black graphics objects such as black lines in graphs, disappear. In this context, the black color *overprinting* feature that was demonstrated in Chapter 8 with the *icecore_piechart_vs12_ai_overprint.eps* file, is of great importance when printing the document on an offset printer. In Illustrator we activate the overprinting feature for black text objects by changing the *Attributes* checkbox to *Overprint Fill*. As indicated previously in Chapter 8, the overprinting feature needs to be treated with care and only applied to black; overprinting of lighter colors, in particular white, results in the complete invisibility of all objects in these colors. Furthermore, overprinting is only possible for vector graphics and not for raster images.

We then list all fonts used in the document by using *Type > Find Font* and eliminating those that are not used in the document or not available on our computer, as these can not be embedded into the PDF file. We use the *Find Next* tool to find the *Wingdings Regular* font, which highlights the list of bullets on the first page of the document. Opening the associated *Paragraph Styles* panel we find that the font is used for the *normal bullets* style, and opening *Bullets and Numbering* in the *Paragraph Styles Options* we find the glyph 131, a black square from *Wingdings Regular* used as a *Bullet Character*.

After completing all these tasks we run the *Preflight Book* tool from the *Book* panel to check all documents included in the book. Again, when all errors listed in the preflight report have been fixed, green bullets appear on the right side of the panel. We can now select *Export Book to PDF* from the panel menu, select the *Adobe PDF (Print)* option, and generate the PDF file which we then save as *sjq_myfirstbook.pdf*. When we generate the PDF file we use an appropriate *Adobe PDF Preset*, having first consulted the printer on this matter. The *PDF/X-1a:2001* format, as an example of an *Adobe PDF Preset*, is a very strict standard for offset printing that does not allow any transparency, embedded interactive features, or the RGB color mode. Large book projects can also be exported chapter by chapter using *File > Export* instead of generating a single PDF file for the entire book, in order to avoid large files. We can activate the *Crop Marks* or *Bleed Marks* checkboxes in the *Export Adobe PDF* dialog box, as well as selecting the correct *Bleed*, if the printing equipment requires PDF files with crop and bleed marks.

Having created the PDF file we open it in Acrobat, adjust the page view, and check the *Output Preview* (which is very similar to the *Separations Preview* in InDesign) using *Advanced > Print Production > Show Print Production Toolbar*. Note that in Chapter 8 we did not set the overprinting feature for black in the figure called *sjq_myfirstbook_fig_2_1_filledcontourplot.eps*, and we therefore have to modify this in Illustrator and update it in InDesign prior to actually generating the PDF file. For the final preflight with Acrobat we use the *Sheetfed offset (CMYK)* profile (or any other suitable profile) which yields a report listing information about the file, including any warnings. These details on the file need to be discussed with the printer, who will also run a preflight and use PDF modification software and fixing routines (such as *Solvero* or *PitStop Pro*) to improve the document and ensure that it perfectly matches the printing process.

Recommended Reading

Johansson K, Lundberg P, Ryberg R (2011) A Guide to Graphic Print Production – Third Edition. Wiley, Hoboken, New Jersey

General Index

Symbole

1-bit arrays   150
3-panel roll-fold brochure   245
4C   267
8-bit color images   151
24-bit true color image   151, 189
64-bit array   152
720p   211
1080p   211
.mat   86

A

A9   22
abstract   17, 37, 201, 237, 239
acknowledgements   236, 239
Action History   246
adapthisteq   188
Add Anchor   178
Add Anchor Point Tool   253
Adding Captions   263
Adobe Bridge   190
Adobe Creative Suite   13, 16
Adobe Illustrator   13, 45, 54, 166, 171,
        175, 177, 183, 225
Adobe InDesign   16, 243, 248, 254,
        256
Adobe InDesign book function   261
Adobe PDF Preset   265, 274
Adobe PDF (Print)   274
Adobe Photoshop   13, 46, 54, 183, 189
Adobe Support   182
Adobe Swatch Exchange   179
Advanced Research Projects Agency
        Network   56
affiliations   239

Align   228, 250
Alignment   257
Align To Baseline Grid   271
alpha channel   192
Amazon   7, 20, 21
Amazon sales rank   22
American Standard Code for
        Information Interchange   53
angled lighting   134
angle measurements   113
anonymous   68
ANSI lumen   211
answer   80
Apple iCloud   74
Apple iWork   11, 15
Apple Keynote   15, 210, 220
Apple Mail   10, 58
Apple Numbers   53, 62
Apple Pages   15, 46, 53, 195, 239, 243,
        254
Apple Preview   45
approach   38
ArcGIS   12
ARPANET   56
array   79
artboard   228
Artboard Tool   233
articles   235
ASCII   53, 73, 130, 194, 198
ASR   22
ASTER   153, 157, 187
Attributes   178, 186, 187
Attributes checkbox   273
Attributes panel   252
Author Guidelines   226, 240, 266
Author Information   249

M. H. Trauth and E. Sillmann, *MATLAB® and Design Recipes for Earth Sciences,*
DOI: 10.1007/978-3-642-32544-1, © Springer-Verlag Berlin Heidelberg 2013

Supplementary Electronic Material

The book's supplementary electronic material (available online through the publisher's website) includes color versions of all figures, recipes with all the MATLAB commands featured in the book, the example data, exported MATLAB graphics, and screenshots of the most important steps involved in processing the graphics.

chapter_1_materials

1_0_figures

afig_1_1.tif
afig_1_2.tif

chapter_2_materials

2_0_figures

afig_2_1.tif
afig_2_2.tif
afig_2_3.tif
afig_2_4.tif
afig_2_5.tif

chapter_3_materials

3_0_figures

afig_3_1.pdf
afig_3_2.tif
afig_3_3.tif
afig_3_4.tif

3_5_icecore_data

icecore_snowaccumulation_data.txt
icecore_temperature_data.ods
icecore_temperature_data.pdf
icecore_temperature_data.txt

chapter_4_materials

4_0_figures

afig_4_1.tif
afig_4_2.tif
afig_4_3.tif
afig_4_4.tif

4_0_recipes

designrecipes_4.m

4_4_data_storage_and_handling

geochem_data.txt

4_6_scripts_and_functions

average.m

chapter_5_materials

5_0_figures

afig_5_1.tiff
afig_5_2.tiff
afig_5_3.tif
afig_5_4.tiff
afig_5_5.tiff
afig_5_6.tiff

M. H. Trauth and E. Sillmann, *MATLAB® and Design Recipes for Earth Sciences*, DOI: 10.1007/978-3-642-32544-1, © Springer-Verlag Berlin Heidelberg 2013

5_0_recipes

designrecipes_5.m

5_2_line_graphs

icecore_lineplot_vs1_matlab.eps
icecore_snowaccumulation_data.txt
icecore_temperature_data.txt

5_3_bar_graphs

icecore_bargraph_vs1_matlab.eps

5_4_pie_charts

icecore_piechart_vs1_matlab.eps
icecore_piechart_vs2_matlab.eps
icecore_piechart_vs3_matlab.jpg

5_5_rose_diagrams

directional_data.txt
directional_rosediagram_vs1_matlab.eps
rose_sqrt.m

5_6_multiplots

multipledata_data.mat
multipledata_multiplot_vs1_matlab.eps

5_7_stratplots

stratigraphiccolumn_data.mat
stratigraphiccolumn_stratplot_vs1_matlab.
   eps
stratigraphiccolumn_stratplot_vs2_matlab.
   eps

chapter_6_materials

6_0_figures

afig_6_1.eps
afig_6_1.tiff
afig_6_2.eps
afig_6_2.tiff
afig_6_3.jpg
afig_6_4.jpg
afig_6_5.jpg
afig_6_6.jpg
afig_6_7.eps
afig_6_7.tiff
afig_6_8.jpg

6_0_recipes

designrecipes_6.m

6_2_gshhs_shorelines

coastline_data.txt
coastline_linegraph_vs1_
   matlab_17703vertices.eps
coastline_linegraph_vs2_
   matlab_8852vertices.eps
coastline_linegraph_vs3_
   matlab_4426vertices.eps
coastline_linegraph_vs4_
   matlab_2213vertices.eps

6_3_etopo2

etopo2_data.txt
etopo2_filledcontourplot_vs1_matlab.eps
etopo2_pseudocolorplot_vs1_matlab.eps
etopo2_pseudocolorplot_vs2_matlab.jpg
etopo2_surfaceplot_vs1_matlab.eps
etopo2_surfaceplot_vs2_matlab.jpg
etopo2_surfaceplotcontours_vs1_matlab.eps
etopo2_surfaceplotcontours_vs2_matlab.jpg
etopo2_surfaceplotlight_vs1_matlab.jpg

6_4_gtopo30

E020N40.DEM
E020N40.DMW
E020N40.GIF
E020N40.HDR
E020N40.PRJ
E020N40.SCH
E020N40.SRC
E020N40.STX
gtopo30_data.tar.gz
gtopo30_filledcontourplot_vs1_matlab.eps
gtopo30_surfaceplotlight_vs1_matlab.jpg

6_5_srtm

srtm_data.hgt
srtm_surfaceplotlight_vs1_matlab.jpg
srtm_surfaceplotlight_vs2_matlab.jpg
srtm_surfaceplotlight_vs3_matlab.jpg

6_6_gridding

normalfault_data.txt

normalfault_filledcontourplot_vs1_matlab.
  eps
normalfault_surfaceplotlight_vs1_matlab.
  jpg

chapter_7_materials
7_0_figures
afig_7_1.tif
afig_7_2.tif
afig_7_3.pdf

7_0_recipes
designrecipes_7.m

7_3_importing_processing_exporting
unconform_image_vs1_original.jpg
unconform_image_vs2_matlab.jpg

7_4_processing_satellite_images
naivasha_data.hdf
naivasha_data.hdf.met

7_5_georeferencing_satellite_images
naivasha_georef_vs1_matlab.jpg
naivasha_georef_vs1_matlab.tif
naivasha_image_vs1_matlab.tif

chapter_8_materials
8_0_figures
afig_8_1.jpg
afig_8_2.jpg
afig_8_3.jpg
afig_8_4.jpg
afig_8_5.jpg
afig_8_6.jpg

8_2_block_diagram
etopo2_surfaceplotlight_vs10_ai_flattened.
  jpg
etopo2_surfaceplotlight_vs11_ai_
  fordarkbackground.jpg
etopo2_surfaceplotlight_vs2_ps_cropped.
  jpg
etopo2_surfaceplotlight_vs3_ai_raw.jpg

etopo2_surfaceplotlight_vs4_ps_isolated.jpg
etopo2_surfaceplotlight_vs5_ps.png
etopo2_surfaceplotlight_vs6_ai_labels.jpg
etopo2_surfaceplotlight_vs7_ai_block.jpg
etopo2_surfaceplotlight_vs7_ai_block.tif
etopo2_surfaceplotlight_vs8_ai_perspective.
  jpg
etopo2_surfaceplotlight_vs9_ps_block.png

8_2_coastline_linegraph
coastline_linegraph_screenshot1.tiff
coastline_linegraph_screenshot2.tiff

8_2_line_graphs
icecore_lineplot_inkscape_screenshot2.jpg
icecore_lineplot_inkscape_screenshot3.jpg
icecore_lineplot_variations_5000px.jpg
icecore_lineplot_vs2_inkscape.jpg
icecore_lineplot_vs3_ai.jpg
icecore_lineplot_vs4_inkscape.jpg
icecore_lineplot_vs5_ai_cmyk.jpg
icecore_lineplot_vs5_ai_fill.jpg
icecore_lineplot_vs6_ai_cmyk.jpg
icecore_lineplot_vs7_ai_greyscale.jpg
icecore_lineplot_vs8_ai_fordarkbackground.
  jpg
icecore_lineplot_vs8_ai_fordarkbackground.
  tif
myfirstposter_example_icecore_bw.jpg
myfirstposter_example_icecore_color.jpg
myfirstposter_example_icecore.jpg

8_2_pie_charts
icecore_piechart_ai_poster1.tiff
icecore_piechart_ai_poster_overprint.tiff
icecore_piechart_swatches.ase
icecore_piechart_vs1_matlab_in_illustrator_
  screenshot.png
icecore_piechart_vs4_ai_transparency.png
icecore_piechart_vs5_ai.jpg
icecore_piechart_vs6_ai.png
icecore_piechart_vs7_ai_smallestfilesize.png
icecore_piechart_vs8_ai_flattened.png
icecore_piechart_vs9_ai_colorzone.png
icecore_piechart_vs10_ai_combined.png
icecore_piechart_vs11_ai_mask.png

icecore_piechart_vs12_ai_
   fordarkbackground.png

8_2_pseudocolor_raster_with_vector

etopo2_pseudocolorplot_vs3_ai.png

8_2_rose_diagram

directional_rosediagram_vs2_ai_lines.png
directional_rosediagram_vs3_ai_fill1.png
directional_rosediagram_vs4_ai_fill2.png

8_3_processing_images

naivasha_georef_vs2_matlab.jpg
naivasha_georef_vs3_ps_rgb_2500px.jpg
naivasha_georef_vs4_ps_rgb_white.jpg
naivasha_georef_vs6_ps_rgb_coord.jpg
naivasha_georef_vs8_ps_rgb_1500px.png
naivasha_georef_vs9_ps_cmyk_670px.png
naivasha_image_vs1_matlab.tif
naivasha_image_vs2_gimp_1772px.tif
naivasha_image_vs3_gimp_autocolor.tif
naivasha_image_vs4_gimp_smallfile.jpg
naivasha_image_vs6_ps_rgb_2500px.jpg
naivasha_image_vs7_ps_cmyk_2500px.jpg

8_4_text

newpaper_vs1.txt

8_5_tables

geochem_data.txt

chapter_9_materials

9_0_figures

afig_9_1.tif
afig_9_2.tif
afig_9_3.tif
afig_9_4.tif
afig_9_5.tif
afig_9_6.tif
afig_9_7.tif

9_4_template

My First Template.otp

9_5_slides

myfirstpresentation_vs1.odp

myfirstpresentation_vs2.odp
myfirstpresentation_vs3.odp
myfirstpresentation_vs4.odp
myfirstpresentation_vs5.odp
myfirstpresentation_vs5.pdf
myfirstpresentation_vs6.ppt
myfirstpresentation_vs7.ppt
myfirstpresentation_vs8.key
myfirstpresentation_vs8.pdf

chapter_10_materials

10_0_figures

afig_10_1.jpg
afig_10_2.jpg
afig_10_3.jpg

10_4_myfirstposter

myfirstposter_1970_1000_empty_ai.jpg
myfirstposter_1970_1000_vs1_ai.jpg
myfirstposter_1970_1000_vs2_ai.jpg
myfirstposter_1970_1000_vs3_ai.jpg
myfirstposter_1970_1000_vs4_ai.jpg
myfirstposter_1970_1000_vs5_ai.jpg

10_4_myfirstposter_materials

etopo2_filledcontourplot_vs1_matlab.eps
etopo2_filledcontourplot_vs2_ai.png
etopo2_filledcontourplot_vs3_ai.png
etopo2_surfaceplotlight_vs5_ps.png
etopo2_surfaceplotlight_vs10_ai_flattened.
   png
geochem_data_table.pdf
geochem_data.txt
icecore_lineplot_vs6_ai_cmyk.jpg
icecore_piechart_vs11_ai_mask.png
myfirstposter_placeholdertext.rtf
naivasha_image_vs7_ps_cmyk_2500px.jpg

chapter_11_materials

11_1_figures

afig_11_1.tif
afig_11_2.tif
afig_11_3.tif

afig_11_4.tif
afig_11_5.tif

11_2_myfirstpaper

myfirstpaper_vs1.odt
myfirstpaper_vs1.pdf
myfirstpaper_vs1.txt

11_3_myfirstleaflet

myfirstleaflet_indd_page_1.jpg
myfirstleaflet_indd_page_2.jpg
myfirstleaflet_indd_screenshot1.png
myfirstleaflet_indd_screenshot2.png
myfirstleaflet_indd_screenshot3.png
myfirstleaflet_indd_screenshot3.tiff
myfirstleaflet_scribus_screenshot1.png
myfirstleaflet_scribus_screenshot2.png

11_3_myfirstleaflet_materials

etopo2_surfaceplotlight_vs5_ps.png
figure_1_chapter_5_neg.tif
figure_1_chapter_5_neg1.tif
figure_4b_chapter_6_cutout.tif
icecore_lineplot_vs8_ai_blue.tif
icecore_lineplot_vs8_ai_fordarkbackground.
    tif
myfirstleaflet.doc
myfirstleaflet.docx
naivasha_image_vs7_ps_cmyk_2500px.jpg
srtm_surfaceplotlight_vs4_ps_cmyk.tif

11_4_myfirstthesis

myfirstthesis_indd_screenshot1.png
myfirstthesis_indd_screenshot2.png
myfirstthesis_indd_screenshot3.png
myfirstthesis_pages_screenshot1.png
myfirstthesis_word_screenshot.png

myfirstthesis_backmatter_jpeg

myfirstthesis_backmatter_1.jpg
myfirstthesis_backmatter_2.jpg
myfirstthesis_backmatter_3.jpg

myfirstthesis_chapter1_jpeg

myfirstthesis_chapter1_1.jpg
myfirstthesis_chapter1_2.jpg
myfirstthesis_chapter1_3.jpg
myfirstthesis_chapter1_4.jpg
myfirstthesis_chapter1_5.jpg
myfirstthesis_chapter1_6.jpg
myfirstthesis_chapter1_7.jpg
myfirstthesis_chapter1_8.jpg

myfirstthesis_chapter2_jpeg

myfirstthesis_chapter2_1.jpg
myfirstthesis_chapter2_2.jpg
myfirstthesis_chapter2_3.jpg
myfirstthesis_chapter2_4.jpg
myfirstthesis_chapter2_5.jpg
myfirstthesis_chapter2_6.jpg
myfirstthesis_chapter2_7.jpg
myfirstthesis_chapter2_8.jpg

myfirstthesis_chapter3_jpeg

myfirstthesis_chapter3_1.jpg
myfirstthesis_chapter3_2.jpg
myfirstthesis_chapter3_3.jpg
myfirstthesis_chapter3_4.jpg
myfirstthesis_chapter3_5.jpg
myfirstthesis_chapter3_6.jpg
myfirstthesis_chapter3_7.jpg
myfirstthesis_chapter3_8.jpg

myfirstthesis_frontmatter_jpeg

myfirstthesis_frontmatter_1.jpg
myfirstthesis_frontmatter_2.jpg
myfirstthesis_frontmatter_3.jpg
myfirstthesis_frontmatter_4.jpg

11_4_myfirstthesis_materials

coastline_linegraph_vs3_
    matlab_4426vertices.eps
directional_rosediagram_vs3_ai_fill1.png
etopo2_filledcontourplot_vs1_matlab.eps
etopo2_pseudocolorplot_vs3_ai.png
etopo2_surfaceplotcontours_vs3_ps.jpg
etopo2_surfaceplotcontours_vs4_
    ps_1780px.jpg
etopo2_surfaceplotlight_vs10_ai_flattened.
    png
icecore_lineplot_vs6_ai_cmyk.jpg
srtm_faults_sat_merged_vs3_ai.jpg
srtm_faults_sat_merged_vs4_ai.jpg

srtm_surfaceplotlight_vs4_ps_cmyk.jpg
table.txt
thesis_backmatter_placeholder_text.docx
thesis_chapter_1_placeholder_text.docx
thesis_frontmatter_placeholder_abstract.
    docx
thesis_frontmatter_placeholder_cover.docx

11_5_myfirstbook

myfirstbook_indd_screenshot1.png
myfirstbook_indd_screenshot2.png
myfirstbook_indd_screenshot3.png
myfirstbook_indd_screenshot4.png
myfirstbook_indd_screenshot5.png

sjq_myfirstbook_jpeg

sjq_myfirstbook_Page_01.jpg
sjq_myfirstbook_Page_02.jpg
sjq_myfirstbook_Page_03.jpg
sjq_myfirstbook_Page_04.jpg
sjq_myfirstbook_Page_05.jpg
sjq_myfirstbook_Page_06.jpg
sjq_myfirstbook_Page_07.jpg
sjq_myfirstbook_Page_08.jpg
sjq_myfirstbook_Page_09.jpg
sjq_myfirstbook_Page_10.jpg
sjq_myfirstbook_Page_11.jpg
sjq_myfirstbook_Page_12.jpg
sjq_myfirstbook_Page_13.jpg
sjq_myfirstbook_Page_14.jpg
sjq_myfirstbook_Page_15.jpg
sjq_myfirstbook_Page_16.jpg
sjq_myfirstbook_Page_17.jpg
sjq_myfirstbook_Page_18.jpg
sjq_myfirstbook_Page_19.jpg
sjq_myfirstbook_Page_20.jpg
sjq_myfirstbook_Page_21.jpg
sjq_myfirstbook_Page_22.jpg
sjq_myfirstbook_Page_23.jpg
sjq_myfirstbook_Page_24.jpg
sjq_myfirstbook_Page_25.jpg
sjq_myfirstbook_Page_26.jpg
sjq_myfirstbook_Page_27.jpg
sjq_myfirstbook_Page_28.jpg
sjq_myfirstbook_Page_29.jpg
sjq_myfirstbook_Page_30.jpg
sjq_myfirstbook_Page_31.jpg
sjq_myfirstbook_Page_32.jpg
sjq_myfirstbook_Page_33.jpg
sjq_myfirstbook_Page_34.jpg
sjq_myfirstbook_Page_35.jpg
sjq_myfirstbook_Page_36.jpg

Lightning Source UK Ltd.
Milton Keynes UK
UKHW02n1140250918
329493UK00002B/49/P

9 783642 325434